Bartolomeo Gastaldi

Lake Habitations and Pre-Historic Remains in the Turbaries and Marl-Beds

Of Northern and Central Italy

Bartolomeo Gastaldi

Lake Habitations and Pre-Historic Remains in the Turbaries and Marl-Beds
Of Northern and Central Italy

ISBN/EAN: 9783337229931

Printed in Europe, USA, Canada, Australia, Japan

Cover: Foto ©berggeist007 / pixelio.de

More available books at **www.hansebooks.com**

Publications of the

Anthropological Society of London.

LAKE HABITATIONS AND PRE-HISTORIC REMAINS IN ITALY.

GASTALDI.

LAKE HABITATIONS

AND

PRE-HISTORIC REMAINS

IN THE

Turbaries and Marl-Beds

OF

NORTHERN AND CENTRAL ITALY.

BY

BARTOLOMEO GASTALDI,
PROFESSOR OF MINERALOGY IN THE COLLEGE OF ENGINEERING

AT

TURIN.

TRANSLATED FROM THE ITALIAN AND EDITED BY
CHARLES HARCOURT CHAMBERS,
M.A., F.R.G.S., F.A.S.L.

LONDON:
PUBLISHED FOR THE ANTHROPOLOGICAL SOCIETY, BY
LONGMAN, GREEN, LONGMAN, AND ROBERTS,
PATERNOSTER ROW.
1865.

PREFACE BY THE EDITOR.

I DO not think that any apology is needed for bringing forward, in an English form, these memoirs of the discovery in Italy of the remains of its early inhabitants, analogous in character to those which have excited so much interest to the north of the Alps.

The work must be taken simply as a sketch of what has been done by Italians in that direction; since M. Desor suggested the suspicion of the existence of such monuments of high antiquity south of the Alps. It consists in great part of reports sent to the author by other men of science who were investigating these remains, and consequently, mainly of facts. Cavaliere Gastaldi himself has introduced them by a short summary; and they are followed by accounts, written by him, of the turbaries in the southern valleys of the Alps, with the inquiry into which he has been most concerned. He has also been so kind as to write for this edition a *précis* of the further discoveries of lake dwellings, made since his memoir was published, which has induced me to alter, with his consent, the title of the work to one more descriptive of its contents, and at the same time more popular.

I may here mention the circumstances which led me to edit the book. During a short stay at Turin, in the spring of 1862, I visited the palace of the Valentino, now used as a college, in which is contained the collection of these antiquities then in process of formation. I was so fortunate as to see it under the guidance of the author, to whose energy it was chiefly due, and who was the more interested in expounding it from learning that I had, not long before, examined the Danish remains of a very similar character, from Jutland, in the Museum of Antiquities at Copenhagen; and, on parting, the author was so kind as to give me his memoir, begging that I would make any use of it I thought proper. I consequently translated it, and in process of time offered it to the Anthropological Society, who decided to publish it, with the concurrence of the author, in their series of anthropological works.

The fact that this mode of inquiry, through the exhumation of buried remains, into the life of the tribes who were external to—perhaps long prior to—the various civilisations which, as it were, in a catena, have conveyed to us the histories and traditions of the earlier world, is a science which has had its origin in the last ten years,—should make us extremely careful in ascribing value to these remains at present. Facts, and an ever larger induction, are of far greater value to us than theories.

If these facts, as they increase, seem to conflict, we must take courage from the reflection of Alexander von Humboldt, "that, in an age of such rapid and brilliant progress as the present, it is a sure criterion of the number and value of the discoveries to be hoped for in any particular science, if, though studied with great assiduity and sagacity, its facts still appear for the most part unconnected, with little mutual relation, or even in some instances, in seeming contradiction to each other."

In the present state, then, of this part of anthropological science, a set of very careful observations, by competent persons, accurately set forth, is of considerable value to those who are carefully working out the problem of the relations between the early tribes of civilised and uncivilised man, and their migrations; and perhaps more in its present form than if moulded into a more regular narrative, and divided into sections and chapters, giving an artificial completeness which, as Bacon says, maketh sciences " stand at a stay," instead of being "added to and supplied further."

With regard to the use of the words "marl" and "marl-bed", although some exception is taken to the form of these words in the Italian, namely, "*marna*" and "*marniera*", by Sig. Strobel and Pigorini, who wish to adopt a special word, "*terra mare*," I have thought it right to keep the word adopted by the author, which is *marna*, the English for which is our ordinary word *marl*.

The word *marna* appears to be of northern origin, or perhaps is an independent form of Pliny's word, *marga*, which he uses in speaking of red and white marls, " quod genus (terræ) vocant " *margam*" (Galli et Britanni), *Nat. Hist.*, xvii, 35. Pliny's form, *marga*, is the connecting link which can be identified with the German *mergel*, our marl (*vide* Zieman's *Middle German Dict.* in voc.); and tracing the word further back we find, as its pedigree, *marag*; Anglo-Sax., *mearg, merg*; Scand., *mergr, medulla, marrow*, akin to the Sanscrit, *mridu tener (tender)* (*vide* Graff's, *Sprachschatz der alt hoch Deutschen sprache*, quoting numerous authorities of early date).

Hence, the meaning appears to be "*soft, marrow-like earth.*" The soft character of much, if not all, marl earth is given by partly decomposed organic remains, which also gives it its value as manure, which property, as Pliny tells us, was well known to the Greeks, and that it was early used in England we know from Chaucer.

> " Se ferd another clerk with astronomie ;
> He walked in the feldes for to prie
> Upon the sterres what ther shuld befalle,
> Til he was in a *marlepit* yfalle."
> *Millere's Tale.*

There seems, then, a perfect propriety in giving the name *marl-bed* as the translation of *marniera* for the locality in which these curious formations—the remains of the past life of ancient tribes—are found; and we need not concern ourselves for the "logical people" of whom Sig. Strobel speaks, who may be tempted to examine the common marls, the character of which is given by myriads of fossil shells and animals.

I shall not here discuss any of the theories that may be held regarding the nations who gave rise to these deposits; what has occurred to me to say I have embodied in one or two notes appended to the report of Signori Strobel and Pigorini upon the marl-beds of Emilia.

C. H. C.

2, *Chesham Place.*

INTRODUCTION TO THE EDITION OF 1862.

STRUCK by the importance which might be attained in Italy, by researches directed to prove the antiquity of the human race, the existence of habitations constructed of wood, anteriorly to the Roman and Etruscan epochs, upon piles in the waters of the lakes and marshes,—to prove, in a word, that our country, like those farther to the north, was inhabited by the races of the ages of *stone* and *bronze*,—I hastened, towards the end of the year 1860, to write a short notice of the discoveries of objects of high antiquity, which had been made up to that time, in Northern and Central Italy.

I hastened, I say, to write that notice,* in order to advise the students of our country of the interest which such researches might have, and to animate them to de-

---

* Published among the *Transactions of the Italian Society of Natural Science at Milan*, with the title of "Cenni su alcune Armi di Pietra e di Bronzo trovate nell' Imolese nelle Marniere del Modenese e del Parmigiano, e nelle Torbiere della Lombardia e del Piemonte."—(Notes upon certain Arms of Stone and Bronze found in the country near Imola, in the Marl-beds of Modena and Parma, and in the Turbaries of Lombardy and Piedmont.)

dicate themselves to it, with the advantage of the fine season; and the haste which was imposed upon me, prevented me from giving attention to illustrating my works with figures representing the objects found.

Persuaded, nevertheless, of the little value which such publications have when they want the graphic representations of the objects to which they refer, I have ever since had the intention of reprinting those remarks, illustrating them by plates; and it is that which I am now doing, inserting in this reprint the fresh discoveries which have taken place, and the information which has come to me in the course of the year 1861. I give to this reprint the title of *Nuovi Cenni su gli Oggetti di Alta Antichita trovati nelle Torbiere e nelle Marniere dell' Italia.*

Part of the designs, which I now publish, were communicated some time back to M. Keller, President of the Society of Antiquaries at Zurich, and were inserted by him, together with an abridgment of my former remarks, in his fourth memorial upon the lake habitations of Switzerland. I am delighted here to be able to make my acknowledgments to him for the honour which he has done to my work; and for his having courteously furnished me with the number containing those designs, which I wanted, that I might unite them with this reprint.

The discoveries made in the turbaries of Piedmont in 1861 are few, and so I shall have little to add to that part of my work which relates to the pile dwellings of Mercurago; but most valuable notices, on the other hand, have been furnished to me, by several of my friends, regarding objects found in other parts of Italy, and especially in the marl-beds of Emilia. These last have been, in the past year, studied by Professor Strobel, with the zeal and care of a passionate naturalist; so that if the publication which I have now in hand, and which contains all the observations made by him, meets with the approbation of the studious, that approbation will be principally due to him.

ENGRAVINGS CONTAINED IN THE WORK.

Fig. 1, p. 16. Piece of stag's horn, with traces left by the instrument used to cut it, from the marl-beds of the neighbourhood of Parma.
Fig. 2, p. 17. Ditto.
Fig. 3, p. 17. Utensil of bone, from the marl-bed of Ponte Nuovo (Modena).
Fig. 4, p. 19. Plan of Conventino di Castione.
Fig. 5, p. 20. Perspective section of the marl-bed at Castione.
Fig. 6, p. 27. Earthenware cup with handle: marl-bed of Castione.
Fig. 7, p. 27. Cup, with appendage perforated by a small hole: marl-bed of Campeggine.
Fig. 8, p. 28. Vase found at Poviglio.
Fig. 9, p. 28. Vase found in the marl-bed of Castellazzo di Fontanellato.
Fig. 10, p. 29. Vase dug up at Enzola.
Fig. 11, p. 29. Vase of the same character and paste: origin unknown.
Fig. 12, p. 30. Internal view of cover of a vase found at Mercurago.
Fig. 13, p. 31. Vase found at Sesto Calende.
Fig. 14, p. 32. Handle of vase from one of the marl-beds.
Fig. 15, p. 33. Handle of a vase from one of the marl-beds.
Fig. 16, p. 36. Needles of bone, from the marl-bed of Campeggine.
Fig. 17, p. 37. Comb of bone, worked in patterns like the Celtic vases, from the marl-bed of Vico-fertile.
Fig. 18, p. 38. Small wheels in bone, from Campeggine.
Fig. 19, p. 40. Hatchet of saussurite, found near Belforte, in the Ligurian Apennine.
Fig. 20, p. 41. Paalstab, in the museum at Parma: derivation unknown.
Fig. 21, p. 42. Paalstab, from the marl earth of Castellazzo.

Fig. 22, p. 42.	Lance-head of bronze, in the Museum of Antiquities at Parma.
Fig. 23, p. 43.	Lance-head of bronze, from the marl-bed of Bargone di Salso.
Fig. 24, p. 44.	Mould in potstone for founding arms in metal: marl-bed of Castelnuovo.
Fig. 25, p. 45.	Spindle-whirls of terra-cotta, of various forms.
Fig. 26, p. 68.	Inferior canine tooth of a bear: marl-bed of Campeggine.
Fig. 27, p. 96.	Hatchet of saussurite, from the Apennine, forty miles from Piacenza.
Fig. 28, p. 98.	Piece of a vase, supposed to be Celtic.
Fig. 29, p. 101.	Illustration of the formation of morainic lakes and turbaries.
Fig. 30, p. 102.	Canoe found in the turbary of Mercurago.
Fig. 31, p. 103.	Plan of a portion of the turbary of Mercurago.
Fig. 32, p. 104.	Horizontal section of a portion of the turbary of Mercurago, containing piles.
Fig. 33, p. 105.	Piece of obsidian from Mexico.
Fig. 34, 35, p. 107.	Vases found in the turbary of Mercurago in 1861, both retaining remains of the cords, tied round the handles, made of twisted ozier.
Fig. 36, p. 111.	Wheel of wood: turbary of Mercurago.
Fig. 37, p. 112.	Wheel of wood of higher workmanship: turbury of Mercurago.

AT THE END OF THE VOLUME.

PLATE I.

Fig. 1. Utensil of bronze: marl-bed of Campeggine.
Fig. 2. Stone Disk (very hard porphoroido-breccia): ditto.
Fig. 3. Flint Arrow-head from the turbary of Bosisio (Lombardy).
Fig. 4. Handle in bone, ornamented, from the marl-bed of Castione.
Fig. 5. Bone-handled bronze Awl, from the marl-bed of Campeggine.
Fig. 6, 7. Bone Instruments: Campeggine.

PLATE II.

Fig. 1. Flint Knife, in the Museum of Turin.
Fig. 2. Stone Weapon, found with the skeletons discovered near Cumarola.
Fig. 3. Bone Arrow-head, from the marl-bed of Campeggine.
Fig. 4. Disk of terra-cotta: marl-bed of Ponte Nuovo, Modena.
Fig. 5. Flint Arrow-heads from the turbary of Mercurago.

ERRATA.

Page 15, bottom line of note,	*for* Thomson	*read* Thomsen.
,, 65, 21 lines *from* bottom (note)	,, aürochs	,, aürochs.
,, 66, 7 ,, ,, ,,	*for* Rhynoceros	,, Rhinoceros.
,, 69, 1 ,, ,, top	,, Berté Eugenio	,, Eugenio Berté.
,, 70, 20 ,, ,, ,,	*dele* Paperini	
,, 76, 7 ,, ,, ,,	*for* shieferkohle	,, schieferkohle.
,, 89, 13 ,, ,, ,,	Annani	,, Anani.
,, 105, 2 ,, ,, bottom	Cerro di las Navajas	,, Cerro de los Navajas.

OBJECTS OF HIGH ANTIQUITY DISCOVERED IN ITALY.

For some years past several different geologists have occupied themselves in investigating new facts, and discovering new elements of discussion which may lead them to the resolution of the question, whether the first men lived at the same time with any of the great animals, of which the most recent fossil fauna is composed—with the *Ursus spelœus*, for instance, the *Elephas primigenius*, etc. The investigations necessary for the resolution of this question, take the study of the remains which are left to us of members of the human race who lived in epochs of which no memory is preserved either in history or tradition, out of the peculiar domain of archæology, and bring it within the sphere of geology.

Consequently most interesting in the double aspect of archæology and geology is the discovery recently made near Abbeville and Amiens, of manufactured flints, mixed up with bones of the *Elephas primigenius,* and of the *Rhinoceros tichorhinus;* and such discoveries as those, which in this case were made, must be not only of high interest, but completely conclusive, if, as appears to be the case, we may exclude every suspicion that these remains of extinct animals could have been already fossil when the waters handling them afresh stratified them in company with the rough products of the industry of the primitive races.*

---

* Later observations, published by the distinguished naturalist, M. Lartet, prove by evidence, that in the south-west of France the human race lived at the same time with the *Rhinoceros tichorhinus,* the *Ursus spelœus,* the *Hyœna spelœa:* see *Nouvelles Recherches sur la coexistence de l'Homme, et des grands Mammifères fossiles réputés caractéristiques de la dernière période géologique.* Now, these species of animals being no longer extant,—in other words, being

These manufactured flints, namely, arrow-heads, lances, hatchets, saws, chips, etc., are certainly the most ancient objects of archæology that we are acquainted with, and mark, if not the appearance of man on the earth, the first results attained to by him in seeking to satisfy his wants.

In some localities of Switzerland, of Germany, of France, of Denmark, and I will say also of Italy, are found exclusively chips of fire-giving flint, and pieces of bone and horn cut into the form of arms and instruments, accompanied by coarse badly baked utensils of earth; in others are found promiscuously arms and utensils of stone and bronze: a metal, the use of which preceded by a long time, as is well known, that of iron; and, lastly, in others are found, mixed with those of stone and bronze, objects of iron.

The use of manufactured stone, that of bronze, and finally that of iron, marking three very distinct steps by which the intellectual unfolding of the human race leapt to the point at which we find ourselves, have induced some archæologists to subdivide the epoch of man into three ages, to which they have given the names of *the age of stone, the age of bronze,* and *the age of iron.* When man learnt how to construct arms and utensils of bronze he certainly did not think of throwing away those which were in his possession of stone, at least for a long time; and probably even after the discovery of iron he could not forget the advantage for many purposes that objects of stone had over objects of bronze; and therefore it is natural that in certain places we should find them mixed together.

The Age of Stone.—In Italy probably belong to the age of stone the many arms of flint, among which was one also of bone, found by Sig. G. Cerchiari upon the hills, which the offshoots of the Apennine form near Imola, and especially in some parts of the parish of Goccianello. Those arms have been described and very well engraved by my friend Doctor

---

fossil species,—it follows that the antiquity of the human race mounts not only beyond the age which we prescribe to it in general, but rises still further to the last great geological epoch; and that epoch having been one of cold, it should not appear extraordinary to us that there should have coexisted with man species of animals formed to live in regions less than temperate.

G. F. Scarabelli, a distinguished geologist, in one of his papers published in the *Annali delle Scienze Naturali di Bologna* in 1850. They consist of about thirty lance- and arrow-heads and of two axes (*couteau-hache*). Of the lance- and arrow-heads some are barely rough cast, the others are of finished execution, a circumstance which, added to another fact of one of such lance-heads being found near a heap of chips of various kinds of flint-stone, and to the greater part of the arms being manufactured from flints found in the neighbourhood, makes one believe that they were made in the country and the very locality where they were found. After the publication of the paper cited above were discovered in the same places small balls of baked clay, with a hole in the middle, little millstones of a garnet-bearing mica-schist, coming in all probability from the valley of Aosta, and some instruments and arms, namely:—

1. A small hammer of diorite of very fine grain, of a blackish blue colour, through which shoot very thin crystals of white felspar. This hammer is sharp on one side and flat on the other; towards the latter end it is perforated by a round hole, of diameter nearly constant (twenty-one millimeters on one side, nineteen on the other), and of wonderful regularity.

2. A cuneiform hatchet of the same stone.

3. A cuneiform hatchet of a stone, probably felspar, very hard and of a beautiful green.

4. Two cuneiform axes of black jasper (Lydian quartz).

These arms, although of the hardest rock, are worked with great fineness.

Arms of the same epoch have lately been discovered by Baron Anca in some caverns of Sicily, and consist of arrows, axes, knives, etc. of volcanic rocks, found mixed confusedly with bones of the stag and hog.* Before him Sig. Forel had found on our coast in similar open caverns, near Mentone, a good number of worked flints in the form of arrow-heads, axes, and disks flattened and sharpened, mixed up with bones

---

* Vide *Bull. de la Soc. Géol. de France*, 1860.

and teeth of the stag, roe, sheep, antelope? ox, horse, hog, bear? wolf, cat, rabbit, with shells, crustaceans, and heaps of fragments of charcoal.* Arrow-heads and chips of flint were also found by my friend Professor Capellini, on the promontories of the Gulf of Spezia. In expectation of the publication of more important discoveries made by the Marquis Carlo Strozzi in Tuscany, I shall here insert, having the permission of my friend Sig. de Mortillet, a letter on the subject by Professor Meneghini, which he received in April 1860:—

"The human bones found in the Cave of the Saints at Monte Argentario have not formed any part of a solid deposit, nor have they been found in the condition of the ordinary accumulations of fossil bones of the caverns. They were scattered superficially in the midst of heaps, which the waters have carried there from a short distance; the arms found with them are points of arrows of the usual form and size. The silex is fire-giving (*piromaca*) light in colour, and does not seem to be of the country. The other bones found together with the human ones are those of the ox, the wild boar, and the rabbit, nor do they present any character by which they are distinguished from the common existing species."

"In another cave in the neighbourhood of Monte Tignoso, near Livorno (Leghorn), Marquis Carlo Strozzi has also found many human bones, and many portions of cranium. The crania are of great dimensions, the parietal bones being rather thin, and in all the teeth are deeply worn, as if from the effect of long and active mastication, it may be of alimentary materials comparatively hard.† This wearing of the teeth not being in proportion with the age. Similar deep wearing away of the teeth has been observed in other human remains found in various parts of the *Maremma*, in all probability similar remains of buried corpses; and it was even noted likewise that the incisor

---

* Forel, *Notice sur les instruments en silex et les ossements trouvés dans les cavernes de Menton.*

† This is the character of the teeth in the crania found in the mounds in Orkney, called Picts' houses; it is also found in skulls in various parts of North and South America.—EDITOR.

teeth both above and below presented the same deep wearing away, which, instead of being oblique, as is usually the case with our race, is perfectly horizontal, a circumstance which has also been noted in the Egyptian mummies.* Together with all the said bones of the Monte Tignoso, the Marquis Strozzi found besides arrow-heads like those mentioned above, heads of lances cut into the shape of olive leaves, and a little hatchet of diorite."

"The fragments of terra cotta, some of them of rude form, have been found within the strata of stone and shells (Panchina), which are most superficial, but not of those which actually form it, except in some rare place, as at the mouth of the *Fossa Calda,* near Campiglia. The same strata of solid *Panchina* containing, together with remains of human industry, numerous spoils of marine animals of species still actually living, are certainly superficial, but raised in some measure above the level of the sea, and irregularly inclined and disconnected."

In 1852 Signor Gabriele Rosa of Iseo, in an article published in the journal *Il Crepuscolo,* printed in Milan, gave an account of the book by M. Boucher de Perthes, *Les Antiquités celtiques et antédiluviennes,* and ended with the announcement that in September 1851, the engineer, Pietro Filippini, in the cuttings made for the construction of the railway from the gate Torre-lunga and the suburb of Santa Eufemia, near Brescia, about two meters below the vegetable soil, found in a stratum of gravel the appearance of a mortuary trench, and in it pieces of charcoal and shards of earthen vases very friable, and fragments of a knife of flint.

We may here remark that arrows of flint and hatchets of different kinds of stone have on several occasions, and at different times, been found in other parts of Lombardy, as well as in the Modenese territory and in Piedmont. These various discoveries made in several regions of northern and central Italy, as well as that of which I have spoken above, made by Signor Anca on the coast of Sicily, gave me reason

---

* Also among the Chinese, Tartars, Esquimaux, and Peruvians.—EDITOR.

to suppose that objects of the same kind might be found in other southern provinces of the kingdom. With such an idea I begged the engineer, Signor de Bosis of Ancona, already known by various writings on geology and the industrial products of the Marches, to be so good as to make some researches directed to this end, and my request could not have been more promptly responded to, for a few days after I received from him a present of seven arrow-heads of flint, which he had collected from country people, "who keep them," Signor de Bosis wrote to me, "to preserve their houses from lightning, believing that the lightning comes down to strike with a similar stone," a superstition which I have found also in Piedmont among the labourers employed in extracting the turf.

The arrow-heads sent to me by Signor de Bosis are in colour white and rose-coloured (it is worth while to mention), of a variety of *piromaca* (fire flint), which presents some analogy with the *piromaca* of which are made the arrow-heads found in Tuscany. Those found in Lombardy, in Modena, and in Piedmont are of milky grey, or blackish brown flint.

I have said above that it appears probable to me that the objects above mentioned belong to the age of stone, because they have been found in considerable quantity, and without any admixture of metal; I shall now indicate much more important discoveries of arms and utensils of the Age of Bronze.

The Age of Bronze.—In a valley called Cumarola, part of the property of Sig. de Gatti, at scarce the distance of a mile from Modena, in the autumn of 1856, were brought to light, in making railway cuttings, about forty human skeletons, buried about three meters below the surface, in the bare earth; these were disposed in two parallel rows, all with their heads turned towards the south, and by their side were arms of bronze and stone. An article* by the celebrated Professor, Sig. Cavedoni, printed in the *Messaggiere di Modena*, published this discovery, and I shall make use of the words

---

* "Archæological account regarding the discoveries of an antique poliandrium, or sepulchral tumulus, of about forty warriors, with their arms."
—*Messaggiere di Modena*, 24th December, 1856, No. 1486.

of the Professor to give an exact description of the particulars observed, and of the objects found :—

"Each one of them (the skeletons) had, on the right side, a lance-head of copper, turned upwards, and on the left side an arrow-head of flint-stone; and besides, some of them had on the right side a cuneiform lance-head of bronze,—some a similar head of the hardest green serpentine, and some over their head a hammer of blackish serpentine, not very hard, and ending on the opposite side in the form of a hatchet; one of them was distinguished by having, on the right side, a lance-head of considerable size and elaborate workmanship, and above its head a pipe of iron,* which, having been broken, appeared full of some substance reduced to minute grains, like semolina."

The author then gives the following description of the objects indicated :—

"1. Lance-head of copper (perhaps with a slight admixture of tin), with a very fine blade, with a rib in the middle; length 0·13 m., breadth 0·07 m., at the base, and these furnished with a prolongation, 0·025 m. in length, with a hole in the middle, the better to fasten it to the end of a staff of wood, of which some traces remain, and which appears to have been of considerable size, and fined down to fit the spearhead at its base.

"2. Lance-point of flint-stone, irregularly cut; length 0·06 m., breadth 0·02 m. with a socket 0·02 m. long.

"3. Cuneiform spearhead of bronze, 0·10 m. long, 0·04 m. broad towards the edge, and 0·02 m. towards the other extremity.

"4. Cuneiform spearhead of green serpentine, of the hardest kind, like the last, but a little shorter.

"5. Hammer of blackish serpentine, covered with small whitish spots, with a circular head, and ending at the other side in form of a hatchet, with an obtuse edge, with a hole in

---

* I have learnt from Sig. di Gatti that it is not certain that this tube is of iron, and on the contrary it is highly improbable that it was; for of all the arms and instruments by which the skeletons were accompanied, not one was found of this metal, but all were undoubtedly of bronze or stone.

it, larger at one side than the other, in which was inserted a short handle of coarse wood, a little longer than the middle finger, of which some trace remains in the earth; the length of this is 0·08 m., breadth 0·05 m., across the middle of the cutting side 0·04 m.

"The skeletons, which were accompanied by the various arms above described, are sufficiently well preserved, and had all their teeth white and entire; certain others of these skeletons must have been exhumed when, about the year 1773, the Via Nuova Giardini was opened, since it cuts across one of the two rows of skeletons now discovered, and probably it was at that time that two arms, like those mentioned above, were found, which are preserved in the museum of the Royal University."

My friend, Professor Döderlein, director of that museum, has been so good as to communicate to me the two arms whence we have been enabled to obtain models. One of them has the form of a crescent moon, from the concave side of which a large and thin appendage proceeds, which must have served to graft the arm into the handle. It is made of green serpentine, darkened by small laminæ of diallage, which may be easily scratched, while it will itself with some difficulty scratch glass, not without, however, receiving scratches in return. The other has the form of a hatchet (*couteau hache*), and is of a felspathic substance (*sausurite* or *petro selce*), of a greenish tint with blackish spots, makes lines easily on glass, resists lime, is so transparent as to show flaws (*scheggiosa*), will melt in a retort, giving a glass of dark green, almost black. (An axe of nephrite, coming from New Zealand, and existing in the mineralogical collection of the school of practical engineering, can easily be scratched with a file.)

I shall now draw attention to one observation, which it has occurred to me to make relatively to the human race who lived in remote times in the Modenese territory,—an observation which may in time appear of little or no importance, but which may, perhaps, still lead the way to new and conclusive discoveries.

Finding myself this summer (1860) in Modena,* and having noticed in the fine museum of anatomy two skulls (coming, if I recollect right, from some cuttings made near Reggio), one of which is, from the inscription it bore, supposed to be of the race of the Zingari, I communicated to Signor de Gatti my regret that the skulls of the forty warriors, discovered at Cumarola, had not been preserved, that we might have seen whether they, peradventure, might not have belonged to the same race.

Some time after, in last September, Signor di Gatti, moved by the generous desire of making some discovery which might prove useful to science, gave his attention to an excavation in continuation of that in which the skeletons, so often mentioned, were found, and discovered a new one; and with extreme courtesy, which I am bound here to publish, not only made a point of being present at the work, so that he might communicate the observations which it occurred to him to make, but forwarded to me all the objects recovered, comprising the skull which, unluckily wanting its facial and its occipital part, leaves it doubtful to what race it has belonged.† I have observed, nevertheless, that this skull, though that of an individual of from twenty-five to thirty

---

* I ought to say that on this occasion I, for the first time, made acquaintance with the works of Ramazzini, Venturi, Brignoli, Cavedoni, etc., and I owe it to the extreme courtesy of Signor Domenico Luppi, the archivist of the corporation of Modena.

† Signor di Gatti, in announcing to me the discovery made by him, wrote to me as follows : " I gave orders that the ground should be excavated with the greatest diligence until the usual indications should be recognised; as it proved after not much labour, the kind of earth was recognised which indicates the interment of a corpse, and then began to be laid bare little bones, we suppose of the feet. I diligently observed whether the bones had, at some former time, been moved; but I saw that this had not happened, from the regularity with which they were placed. The length of the skeleton appearing to me out of the common, I measured the length, and found it 1·91 m.; the bones of the skeleton continued to be taken out with the hand, and at the distance of 0·57 m. from the skull, towards the feet, we found a quantity of greenish earth; this green earth having been removed with every precaution, there was presented an object, in length 0·08 m., in breadth 0·065 m., which those standing by adjudged to be a decoration, which must have been placed on the breast of the corpse. From the investigations made we may gather that this decoration was composed of tissue, and had

years of age, has the frontal (*fronto-frontale*) suture very apparent,* it happened then that reading the "Statement of the Natural History of the extensive Estates of the Signori Brignoli and Rezzi," I saw that the authors mention two skulls (the very same which I have alluded to as in the Anatomical Museum at Modena), found about four meters deep, at Caudelbosco, near Reggio. "One of these skulls," say the authors, "appears to have belonged to a Zingaro (Cinganus), the other certainly belonged to an individual of the Caucasian race; but this last, which is that of an adult (of the age of from thirty to thirty-five years), presents the peculiarity of having throughout apparent the frontal (*fronto-frontale*) suture, and we know that they join, so that ordinarily, after the seventh year, one cannot perceive them, though there are sufficiently rare instances in which this property may have been observed by some anatomists."

Here, then, we have two skulls; one found near Reggio, the other at Cumarola, near Modena, both found at about four meters depth,—one, of the Caucasian race, the other of uncertain race, both presenting the unusual peculiarity of showing, in a marked manner, the frontal suture. It seems to me that, in the coincidence of these two facts, there is something more than accident.†

Marl Beds.—In the provinces of Parma, of Reggio, and of

---

the form of an oblong radius (*raggio oblungo*), with little tubular bodies (threads of copper) imprisoned in the tissue; and in the middle of the opening there was another small object still longer. We tried in every way to preserve the decoration I have described entire, but it was impossible; the tissue which kept it together was destroyed, and the earth on which it rested, under the influence of the air, became reduced into minute fragments. I have, however, carefully collected the tubes and the object which was in the centre. Proceeding with our discoveries, we found on the right side the usual lance of metal (bronze), and lower still, an object also of metal, in form of a wedge. At the time you spoke to me of the skeletons found in 1856, at Cumarola, I felt that it was a matter of regret that the skulls were not preserved; I however procured that the bones I then discovered should be collected with the greatest diligence, and carried them all to Modena, to place them at your disposal."

* *Saggio di Storia Naturale*, etc.

† Having gone expressly to Modena in the course of December (1860), I have been enabled to ascertain that the two skulls, preserved in the Anatomical Museum of that city, are in fact those which the Signori Brignoli and

Modena, there are to be found, in some situations, considerable deposits of an earth of nature and colour most various, being very rich in animal substance. These earths are much appreciated, and are utilised to improve and manure the meadows; in the country they are known by the name of *marne* (marl); Venturi has called them cemetery earths (*terre cimiteriale*). In the marl-beds are found earthenware, utensils, coins, etc., of the Roman epoch; arms, utensils, earthenware, of a more ancient epoch; and finally, human bones, and those of the horse, the ox, the stag, the pig,—all often mixed with ashes, charcoal, carbonised cereals, etc.

Here I shall transcribe the analysis of one of these earths, due to Professor Merosi, and published by Venturi in his *Storia di Scandiano*.

Four hundred grains of cemetery earth, very loose, *i. e.*, sandy, and containing alum, have given—

Water absorbed	15
Loose rock and gravel, part flinty, part calcareous	30
Vegetable fibre, not decomposed	9
Fine flinty sand	196
Human bone, not decomposed	23
The remaining material, minutely divided and filtered, has given—	
Carbonate of lime	12
Carbonate of magnesia	4
Matter, chiefly animal, destructible by heat	18
Silex	29
Alumina	31
Oxide of iron	4
Soluble material, principally phosphate or sulphate of potash, and vegeto-animal extract	8
Sulphate of lime	3
Phosphate of lime	17
Total ...	399

Rezzi speak of in their *Statement of Natural History*; and I was shown, by the celebrated Cavaliere Gaddi, Director of that Museum, a third cranium, of an adult with the frontal suture. This skull was discovered in 1858 by the Professor Signor Cesare Costa, in excavating a well in the heart of Modena, and lay in a Roman sarcophagus, with a lid closed and leaded.

Many instances similar to the above are known. Welcker states that the proportion of German skulls in which the frontal suture is present, is at least one in ten; Thurnam states, that the proportion in the English and Irish is about as frequent; each alleges that in French skulls, from the Paris catacombs, it is one in eleven.—Thurnam in *Mem. Anth. Soc. London*, vol. i, p. 155.—ED.

Venturi, as I have said, calls such deposits cemetery earths, and divides them into three categories. In the first he comprises the most ancient earths, those which he supposes to have been either burial places, in which were interred the warriors of the Boii (a Celtic nation) who perished in battle, or else remains of the funeral piles and sacrifices of the same epoch: in the second he comprises those which are rich in the sepulchral remains of the Roman epoch: and finally, in the third, the heaps resulting from the ruins of old fabrics destroyed by time or burning. Cavedoni, in an article on a sword of bronze, found in the *marl bed* of Marano, and by him believed to be Roman,* gives his adhesion to one of the opinions thrown out by Venturi, and considers such *marl* as the remains of piles and sacrifices, noticing, very appositely, on the subject, that " as the material composing the *marl-bed* (that observed by him) consists of strata distinctly accumulated one over another successively,—certainly not formed at once, and disorderly, since each one of those layers is wont to be covered by a row of stones and fragments of brick, of terracotta, and of natural earth,—one cannot consider such marl-beds as 'heaps of corpses dead in battle, and burnt pillage of the Galli Boii,' as Venturi supposes."

Following indications kindly given me by Professor Döderlein, I visited the *marl-bed* of Casin-Albo, and that of Ponte Nuovo, near Sassuolo.

At the marl-bed of Casin-Albo I saw a deposit generally formed of horizontal beds of a brownish clay, containing many fragments of Roman earthenware, and in particular of sepulchral tiles, accompanied by human bones and those of quadrupeds, as well as rounded flints of considerable size; the stratification is at many points not very apparent, in others it is more so, and indicated by thin layers of rounded flints; in other places the clay contains much carbon, and becomes black; finally, in other places are found marsh shells and land shells. The *marl-bed* is about three meters deep.

Observations were also made on one near the last, at Ponte

---

* Di un' antica spada Romana rinvenuta nella così detta marna di Marano.—*Indicatore Modenese*, anno 2, No. 18.

Nuovo. There, however, we did not meet with coarse Roman earthenware as at Casin-Albo, but fragments of earthen vases, black or reddish, imbedding small grains of quartz. Here, too, the deposit is an alternation of rounded flints, blackish clay, bones, carbon, and rubbish of pottery, the whole broken up into minute portions.

If I might be allowed to draw some conclusion from the little that I saw, I should say that neither Venturi nor Cavedoni have noted an important fact; they have not observed, namely, that in the majority of the *marl-beds* the objects discovered have been displaced and rehandled by water, and are no longer in the site which they occupied at the time of interment.

It appears to me that the discovery made in 1856, at Cumarola, of the forty warriors buried in the same spot, may up to a certain point put us on the way to reach the interpretation of the origin of the *marl* deposits. I shall premise that all the *marl-beds* of the territories of Reggio, Parma, and Modena, of which mention is made by Venturi and Cavedoni, are situated in proximity to torrents (an observation which was made to me by Professor Döderlein).

It is known that the regions in which we meet with these deposits have, in past ages, been subject to devastating inundations, caused on the one hand by the overflowing of the Po, and on the other by the bursting from their banks of the torrents which, descending from the Apennine, discharge their streams into the Valle Padana. Finally, we know that some of these *marl-beds* are stratified.

That being granted, if we here imagine that kind of burial-place discovered at Cumarola in 1856, and another similar one situated at a lower level; in other words, if we imagine a superficial stratum of ashes, carbon, and broken bones—in a word, a stratum, a bed of what has been left at the meetings or funeral feasts of the aboriginal races, cut up, washed, and re-arranged by the waters inundating the surrounding country, we shall easily understand that there would result from it a deposit very analogous to the circumstances of the *marl heaps*, a deposit, that is to say, stratified, consisting of sand, clay, and flints, and containing ashes, bones, objects of human industry, etc.; that if in more recent times other inundations should have come successively, so as to deposit new strata, containing, it may be, earthenware and other utensils of the Roman or post Roman age, the deposit which would result from it would be a *marl-bed* of the most complicated and perfect kind.

If from the little, I repeat, which I have seen, and which I have been able to gather from the writers who before me have spoken about these most curious earths, I may be allowed to draw general deductions, it is my opinion that the majority of the *marl-beds* are remains of the burial-grounds of the Romans, and remains of the cemeteries, funeral piles, and better, of the feasting places (kjökkenmödding), meeting places, or stations of the ages of bronze, for the most part remodelled by water.

I have said that the *marl-beds* are remains of Roman burying-places, and the fragments of the well-known tiles with which the Romans used to cover their sepulchral houses, the lamps, the tear bottles, the coins, and other utensils of that epoch which are very frequently met with in some *marl-beds*, make it unnecessary to prove it.

I have said also they are probably remains of cemeteries and pyres of the epoch of bronze, and that, because not unfrequently there are found in them, mixed with ashes, carbon and human bones :—

1. Arms and instruments of stone and especially darts of flint, accompanied by arms of bronze, that is to say, swords, lance-heads, cuneiform hatchets, axes with a lateral reflexion (paal-stab).

2. Vases roughly worked, badly baked, black or dark red in colour, made of clay embedding granules of quartz,* as well as certain balls of terra cotta pierced with a hole, called

---

* In the *marl-beds* of Modena the earthenware, and principally the pre-Roman, comes out always in fragments, but having seen in the *Storia di Scandiano* of Venturi that in the Museum at Parma there existed some entire vases from the *marl-beds* of the neighbourhood, I visited that museum in the course of December 1860, and thanks to the extreme courtesy of the director, Cav. Lopez, I found there a series of vases (of some of which I shall presently give the form), which, in workmanship, material, or figure,

by Cavedoni *fusajuole* (spindle-whirls).* (Cavedoni, " *Di un' antica spada Romana rinvenuta nella cosi detta Marna di Marano*"—*Indicatore Modenese*, an. 2, No. 18) which are also found, as I have said, in this country, near Imola:

All objects identical with those which in Switzerland, Germany, Denmark, and in other countries, are characteristic of the pre-Roman age, and particularly of that of bronze.

I have said that it is also probable that the *marl-beds* are remains of feasting places, meeting places or stations; not only on account of the large number of bones of the ox, stag, pig, etc., which are found in them, but particularly because a large proportion of these bones, those, namely, which are large and hollow, appear as if always broken at one end, in almost exactly the same way in which the aborigines of other countries are wont, both in Greenland and Lapland, to break the hollow bones with the view of extracting the marrow.†

Here I may give what M. Morlot writes on the subject in his *Studii Geologico-Archeologici* (Lausanne, 1860). Speaking of the objects found in the kjökkenmödding: "One circumstance to point out is, that all the solid bones, without hollow parts, of quadrupeds, are entire; while those which are hollow we find, with scarcely an exception, broken, often showing the mark of the blow by which they have been opened. The population was evidently greedy of marrow, which it took wherever it could find it, either to eat or to employ it with the brain in the preparation of skins, as the North American savages do at the present day."

Enumerating afterwards the objects found in the middle of the palisades (the lake habitations) of the Swiss lakes, he subjoins:—"The pile system of Mooseedorf has furnished an

---

in a word, by that union of characteristics which in natural history is known by the term *facies*, are assimilated to those of the bronze age, found in the turbaries of Mercurago and San Martino, of which we shall speak hereafter.

* It is singular that the name *fusajuole* (spindle whirls) given to these balls by Cavedoni, corresponds so exactly with *pesons de fuseau* (spindle weights), by which they are spoken of by Troyon.—*Habitations lacustres des temps anciens et modernes*, by Frédéric Troyon, Lausanne, 1860.

† This remark was made to me about the Jutland heaps in 1857, by Professor Thomson of Copenhagen, *vide Anthrop. Review*, vol. ii, p. 60, 1864.—ED.

abundance of broken bones of animals: we see that here, as in the north, man has opened all the hollow bones to extract the marrow."

Professor Döderlein for many years has observed that the long bones found in the *marl-beds* present traces of blows of hatchets, a fact which was confirmed to me by the examination of a considerable quantity of bones forwarded to me by the same Professor Döderlein, and by Signor Leopoldo Grassi, archpriest of the parish of Sassuolo, a passionate collector of the natural curiosities of his country.

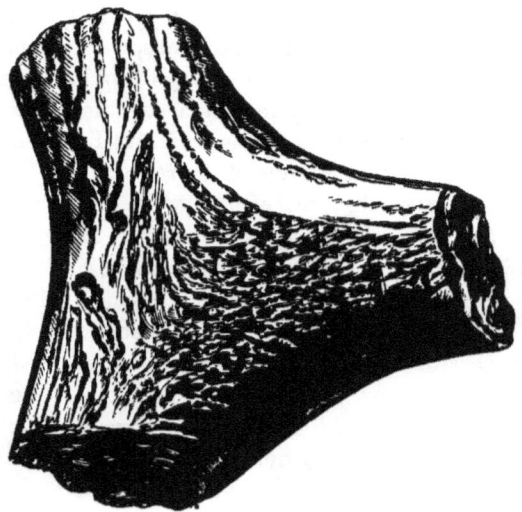

Fig. 1.

In company with the bones are found in the *marl-beds* great quantities of pieces of stags' horn, which almost always bear traces of the labour of man; some of these pieces or secondary branches have been separated from the principal trunk by means of a cutting instrument; however, the cut has rarely effected the complete separation of the piece from the stem or principal trunk, and almost always it has only penetrated far enough to allow of the subsequent separation by breaking the part uncut, whereas the pieces are rare that have the face of separation clean; in some the splinter of fracture is wanting, in others it is more evident. The branches of horn, of considerable size, have generally been detached not with a cut, but evidently with repeated blows of some trenchant instrument. I owe it to the courtesy of Professor Döderlein to be able to give an engraving of a piece of large horn, cut in this way.

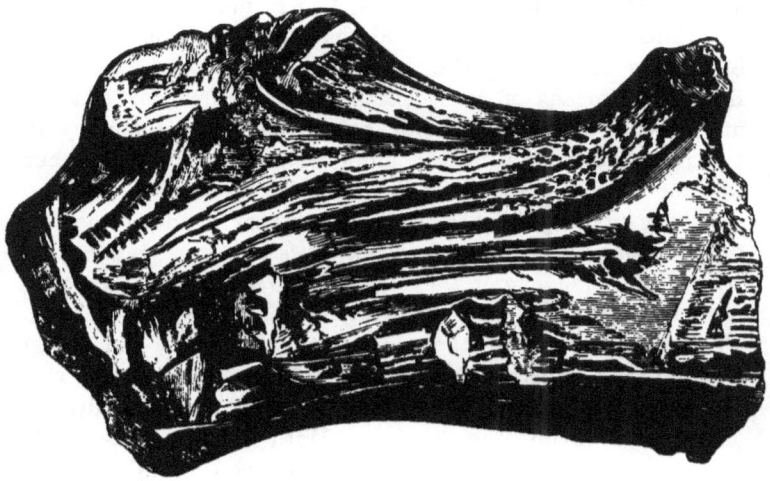

Fig. 2.

This specimen comes from the *marl-beds* of the Modenese, and belongs to the Mineralogical Museum of Modena. Exam-

Fig. 3.

ples not less interesting have been given me by Archpriest Grassi, from which I give one (fig. 3); it is a piece of stag's horn, cut into the form of a tube, terminating at one end in an expansion in the form of a disk, pierced with several holes, the lines, hacks, and repetitions of which this instrument give traces, demonstrate clearly the difficulty which the artificer experienced in working on a material so hard and so impervious to the instruments which he had at his disposal.

I think it unnecessary to add that the operations, whether of cutting or hacking the bones or horns, have been performed by means of instruments of hard stone; of which I am convinced by not having succeeded in producing with the instruments of bronze coming from the *marl-beds* hacks like those which are seen in these bones or horns.

In the notes published in the past year, I ended the few observations made by me on the *marl-beds*, by expressing the wish that some one of the naturalists resident in Parma or Modena would take up the study of those curious deposits. Professor Strobel, of the University of Parma, has amply responded to my call, giving himself up to the study with all the activity and intelligence of which he had already given proof in his other labours. With him was afterwards associated a distinguished young man, Luigi Pigorini, student at the Museum of Antiquities of Parma; and I consider myself fortunate in being able to insert here their observations, as well as those of Signor Gramizzi:—

REPORT OF SIGNOR
PELLEGRINO STROBEL

ON THE MARL-BED OF CONVENTINO DI CASTIONE, AUGUST 1861.

The Conventino, a kind of convent castle, rises from a little hill, a quarter of a mile north from Borgo San Donnino, of which commune it forms part; the summit of the hill stands about three meters above the level of the surrounding plain of the Po, and its base of two hectometers, or perhaps more in extent, has an elliptical form, with its larger diameter from S.S.W. to N.N.E., as appears from the plan with which I am favoured by the engineer Signor Eugenio Berté.

The space of ground where, thirty years ago, the *marl earth* (*terra marna*) was discovered, and from which they have excavated it up to the present time, is situate to the east, and at the base of the mound, and occupies an extent of about thirty-seven *ares*;* it has the form of a lengthened quadrilateral, very narrow at one of the two extremities, and in that place

Fig. 4. Ground Plan (scale $\frac{1}{1000}$) of Conventino di Castione.

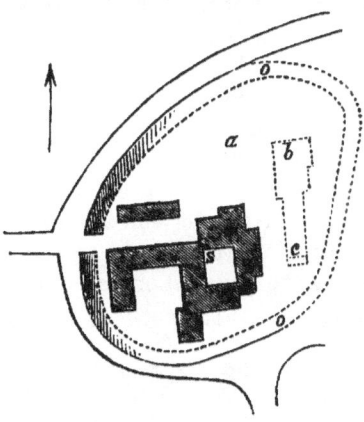

a. Part in which, as they say, the *marl earth* is not found. *b*. Old Excavation, now filled in and cultivated. *c*. Actual Excavation. *s*. A point at which is found the marl earth, by boring with an auger, at about 2 meters down. *o*. Trench.

where at the present time the marl earth is being taken out, one continually sees part of the great system of piles (*pallafitta*), while the rest of it is covered and hidden by cultivation; the marl earth has 2·50m. of elevation (*potenza*) above the head of the piles.

Here is as near as possible the order in which the beds are deposited which form the mound of Conventino, beginning with its summit, or from the bases of the building:—

---

* The "are" is, in the French and Italian metrical system, the square décamètre (10 mètres) and unit of surface, and = 119·6033 square yards, or 0·0247 acres. It may, perhaps, be as well to add that the mètre, or unit of length, = 39·3708 inches, or 3·2809 feet; the hectomètre (100 mètres) = 328·09 feet, or 109·3633 yards.

1. Transported earth - - - 2·00 m.
2. Marl earth (*terra marna*) - - - 2·50
3. Blackish earth, woody—calcareous - - 1·00
4. Greenish ashy earth, calcareous, somewhat less than the preceding - - -

In the second stratum, *i.e.*, in the marl earth, are found deposits of ashes, with charcoal and broken pieces of earthenware, bones of oxen, teeth of the horse and the wild boar, horns of the stag, a certain amount of the fossil *Cardium* and *Unio*, which are more abundant in the third stratum. I was assured besides that from the second stratum, and from the upper part of it, were disinterred two skeletons entire, and lying on their backs. Among the objects of human industry there are found in that stratum pieces of earthenware of many sorts, the sides of which are more or less thick, in colour either black or reddish; fragments of millstone of mica-schist, bearing garnet, scoriæ, etc. The third stratum we may call that of the palisade, because that is the object it contains, which is most inte-

Fig. 5. Perspective Section of Excavation of Marl Earth.

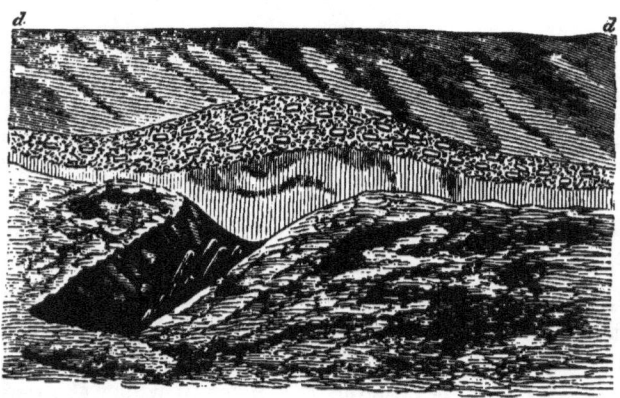

d d. Level at two meters below the base of the buildings. *e.* Bed of ashes, shards of earthenware, etc. *f.* Palisade (*palafitta*). The heads of the piles are found at 2·55 meters from the level *d d*, and at 4·56 meters below the top of the hill.

resting and characteristic; the piles, as far as appears, of elm and chestnut, are about two meters in length, and in breadth 0·12m. to 0·15m., disposed one touching the other, and by a pressure or force exerted from W.S.W. inclined to E.N.E., forming with the horizon an angle of 20°. One of the diggers assured me that at the top of these piles were found great planks, but at present there is not one to be seen; instead of this some of the stakes are carbonised at their ends. When they were vertical the superficies of the palisade (*palafitta*) must have resembled a pavement of wood, such as one sees in some northern countries, on which perhaps rested a platform of planks. The palafitta has been found up to the present time under the marl earth, wherever it has been excavated, that is on an extent of thirty-seven ares, and perhaps it will be found under the whole base of the mound, except towards the north, where, as I have ascertained, the marl earth is not to be found, while a trial with the auger showed it to exist under the summit. From among the piles were brought to light various specimens of the *Unio*, one of which was so well preserved as to seem as if just abandoned by the mollusc, the ligament and the mother-o'-pearl (*madreperla*) were fresh; the last, however, at the contact with the air turned the colour of lead. From the piles were extracted, besides bones, roots, seeds, and acorns entire, a cake of terra cotta with a hole in the middle (a sort of large spindle-whirl), and fragments of earthenware. Other objects have in like manner been found by Cavaliere Giuseppe Ugolotti-Manarini and Signor Ugolotti Romualdo of Castione in these excavations, without however any indication of what stratum they were recovered from; they are horns of gigantic stags, of roe deer, boars' tusks, a cup of blackish terra cotta, little spindle-whirls (*fusajole*), a haft of worked bone, a branch of stag's horn bored with a hole and polished, daggers and a knife, a hatchet of bronze, etc.; a dagger of bronze found there has been stolen from the proprietor.

REPORT OF SIGNOR

MASSIMILIANO GRAMIZZI
ON THE MARL-BED OF SALSO MAGGIORE.

The *marl-bed* in question is situated in the borough of Salso Maggiore, four miles distant from Borgo San Donnino, to the right of the road which leads from Borgo to Salso. It is upon a hill; I have not been able precisely to ascertain its extent, the marl earth being found in different positions; the extent of the excavation from which the majority of the earth is extracted is about 400 meters square, the depth to which it reaches is 1·50m., which is visible in the place where at present excavation is made. In some situations the *marna* is found on the surface of the ground, in others under a sandy stratum of 0·50m. in thickness. There are found at various depths charcoal, bones of men, many bones of animals; indeed of these there are so great quantities that the excavators, in winter, collect them and burn them to warm themselves. Elsewhere are excavated fragments of earthen vases, pins and lance-points of bronze, and many coins, nearly all about the size of our *centessimi*. At 300 meters distance from the marl-bed runs the torrent Gerra. With regard to traditions, I have found the following in a little work of a certain Professor Ghiozzi Borghigiano, who, wishing to prove that Borgo San Donnino had been built on the ruins of the ancient Fidentia, thus expresses himself:—"To give weight to the opinions I have expressed, I bring to the recollection of the courteous reader, that on the hills of Bargone, situate in the quarter of Valle, at some little distance to the right of the torrent Stirone, and four miles from Borgo San Donnino, there is a place commonly called 'Terra Marna,'* on the road to Salso Maggiore, which would imply a sediment of the sea; but this term seems improperly applied, since it is proved that in such a place the nations met to make their sacrifices in the most remote pagan times, that is, the Etruscans, the Galli Anani and Galli Boi, and afterwards the Romans who inhabited our plains; since there are found in great plenty in this mound of earth, rendered fertile by the slaughtered victims, and in the strata of ashes, as it were saponaceous, which are in the midst of it, all the instruments employed in sacrifice, such as axes (single and double), short swords, knives, daggers, long skewers made of wood and metal, clay vases in which they burnt

---

* For discussion of this word and its origin, see Preface by Editor.

perfumes, besides various interesting objects in gold, laid bare in such a manner that they have exported more than 4000 cubic *quadretti* of this earth, whence one may suppose without being far wrong, that in that vicinity there was a considerable population or city, which there, for many centuries, gave worship and honour to the false divinities."

REPORT OF SIGNORS
STROBEL AND PIGORINI
ON THE MARL-BEDS OF EMILIA.

Now that scholastic occupations and the inclemency of the season, make it, if not altogether impossible, at least sufficiently difficult and laborious to continue our researches on the "*terra mare*,"* we may take advantage of this pause to pass in rapid review the objects of nature and art gathered in it, and then to make a short *résumé* of the observations made upon those earths. In this report, which has it in its power to direct us according to its will, it was our intention only to set forth the bare facts, leaving to others for the present the duty

---

* This is the name which our countrymen give to the azotized earths, called by Venturi cemetery (*cimiteriali*), and by others, improperly, marl earths (*terra marne*). The first of these terms, that of Venturi, is too special, and does not embrace the various sorts of those deposits; and the second conduces to error as generally, as well as among us,—by marl is intended chalky clay (*il calcare argilloso*). Wherefore, not to perpetuate this source of *equivoque*, and not to have in the future again the annoyance of inducing certain *logical* people to search in our blue marls (*marne azzurre*) for the objects of the *mariere*, we adopt the above vulgar word, which, if it has not the advantage of expressing an idea, has not, at least, the disadvantage of offering a *false* one. As regards the component parts of these earths, and the use made of them in agriculture, we can only here refer to what has been said on the subject in "Cenni su alcune armi di pietra e di bronzo," etc., inserted in the *Transactions of the Society of Natural Science at Milan*, vol. iii, p. 11, 1861.

I have given the origin of *marna* in the preface. It is clear that *terra mare* is a word referring these deposits to *marshiness*, being connected with *marese*, a *fenny place*, which, it is clear, has nothing to do with the origin of the word; as *marl* means earth containing organic matter in various stages of decay, I have chosen to apply it as the only reasonable translation of the word *marna* or *marniera*, the latter undoubtedly meaning a *marl-bed*, and being the word used by the author.—EDITOR.

of drawing conclusions, physical or archæological. But we confess that in setting forth these facts in a certain order, we have not been able to resist the temptation of allowing our minds to run to some conjectures—to advance some hypotheses. Pardon this boldness of ours, the effect of the natural inclination of any one devoted to these studies. However, we do not hold much to our deductions, and we shall ever be ready to retract them, where further facts and more extended observations, diligent and minute, which perhaps in the future it may be possible to institute, shall have persuaded us that we are in error.

For greater distinctness we have divided our statement into different heads; let us begin with what regards the

WORKS OF MAN,

a head which we may sub-divide into five paragraphs, which treat respectively of the habitations, the vessels, the utensils, the arms, and the things of uncertain use.

I. HABITATIONS.

Of the different sorts of habitations of the people of the marl-beds, without doubt the most important are the palisades (*palafitte*). Up to the present time they have only been discovered at the Conventino of Castione, where, as far as we know, no one has undertaken further works of excavation, and from which place one of us, in the course of the summer, transmitted a detailed report; we can only refer to it, not having found anything to add to it, except the expression of our longing for the continuation of these excavations,—so very interesting, and under a scientific superintendance. From the aforesaid report it results that the people, who are conventionally called *Celts*, were accustomed to use in such constructions in our marshes, as well as in the lakes of Piedmont and Switzerland, small piles, with the difference that with us, to reduce them to that size, they did not split up the trees lengthways twice or four times, as the trees were sufficiently thin in themselves. Naturally in Switzerland they brought into use the firs, beeches, oaks, and birches, which abound there; while here they had recourse to the common elm and chestnut.

According to Strabo (B. IV), the houses of the Celts were round, constructed of planks and hurdles (*graticcie*), and covered with a roof of thatch (*stoppia*, stubble). In France and England, according to Caumont,[*] they were also often rectangular, and had a foundation of unmortared wall (*mura a secco*),[†] often lower than the surrounding level, whether to defend it from the inclemency of the weather, or perhaps not to give to a not very solid wall too high and dangerous an elevation. But, besides, in France, as we are informed by the same author,[‡] certain inequalities of ground, some traces, little visible, of rough foundations indicate, to the archæologist, the site of the Celtic habitations.

And this is exactly what has been observed in some of our *marl-beds*. Signor Giuseppe Sandri, proprietor of a marl-bed at Madregolo, assures us that, making excavations in the past year, he met, at a considerable depth, walls constructed of clay. In the marl-bed of S. Polo are to be seen the vestiges of an unmortared wall. From the *marl-bed* of Vico-fertile was excavated (we give it on the authority of the towns-people), only in the past year, a beam, at about a meter in depth, which, perhaps, belonged to such a Celtic habitation.

II. Vessels.

These are either of terra cotta or stone. The first, or the earthenware vessels of which we shall discourse here, are comprised by us under the name of

Pottery (*Stoviglie*).

We shall here only occupy ourselves with those vessels of

---

[*] *Cours d'Antiquités Monumentales professé à Caen*, tome i, première partie, *Ere Celtique*, p. 157.

[†] This is the principle of the Norwegian houses of the commoner sort,—a pile of stones, without mortar, carefully built, often lowered into the surrounding land, on this a wooden house of logs or planks; also, in Siam, *vide* Bowring,; also in Japan, *vide* Japanese drawings.—Editor.

[‡] *Loc. cit.*, p. 156.

earth, which are kneaded of sandy earth with grains of quartz, which have not been baked in an oven, and therefore are little indurated, and present a fracture altogether peculiar to themselves; such as we see extremely well figured and coloured in the Celtic antiquities of Caumont,* and are distinguished by the archæologists by the name of *Celtic*. And here we may lay it down, once for all, that in this exposition we shall not take account of the objects of art, although found in excavating the *marl earth*, whenever we are sure that they are the product of Roman industry, or of later date.

Although the paste of this pottery, as we have said, presents an appearance peculiar to itself, yet we can distinguish differences corresponding unquestionably to the various uses they were to serve. The larger vases—intended, in all probability, for cooking, and to contain liquids and cereals, and perhaps also the ashes of the dead—are roughly kneaded; and here the grains of sand are larger and more visible; the colour of the paste is blackish ash colour inside, and reddish in the external coats; the sides have a considerable thickness (0·018 m.),† the base still greater (0·025 m.), this, which is smooth, presents a diameter of from 17 to 24 centimeters. As is natural, they show no traces of varnish either externally or internally; the colour of the superficies is a clear reddish tint, which now tends to ash colour and now to yellow, which may give presumption of the colour which we may conjecture the original paste of these vases to have been: no whole one was found in our *marl-beds*, but a profusion of their shards.

Other vases are made with greater care and precision, and of finer paste, so that one at times has difficulty in recognising the siliceous granules in them; the sides are thin, even very thin (0·002 m.), and the use to which they were destined must therefore have been somewhat different from that of the dishes above alluded to, and their forms of plates (*scodelle*), cups (*coppe*),

---

* *Ibid.*, p. 156.
† The *millimètre* = 0·03937 inches; the *centimètre* = 0·39371 inches.

basins (*tazze*), bottles (*oricanni*), indicate it. Their colour inclines to blackish for the most part, and is sometimes ash-coloured, or else yellowish, with innumerable black specks.

Fig. 6.

The tint of the paste is analogous to the tint of the superficies, as we have described it. Of the thousands of shards

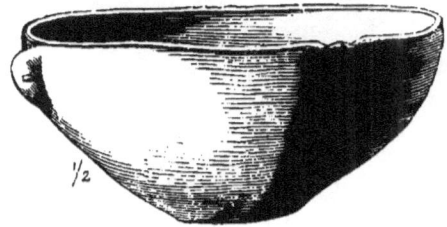

Fig. 7.

of this kind of pottery, which enter into the composition of the *marl-beds*, only sixteen vases, as far as we can learn, have been taken out entire; two of these are represented above; nine belong to the local Royal Museum of Antiquities, and five are in our own possession.

Between these two extremes of coarse and fine vases, now described, all intervening kinds are found, which, in colour of the surface and the paste, belong to one class or the other as they incline to each respectively. In the midst of the rubbish of these intermediate kinds of pottery were found some pieces made of a paste of red earth, of dark brick-colour, containing, still, sandy grains, and which appears to have been baked in a hotter fire.

It is only the fine and bright earthenware, which presents the appearance of a varnish, which, however, does not alter its colour, and appears to belong to the substance itself of the vase; sometimes, however, it disappears after washing.

Now, let us proceed to the analysis of the form of the earthenware, as far as we can gather it from the few entire vases, in part figured here, and from the larger shards; in some a ridge (*carena*, keel), more or less raised, and sharp rims horizontally on the side, as in the vase found at Poviglio (fig. 8); and the vases, so rimmed, are rather shallow

Fig. 8.

in comparison with their diameter, for the most part small, and with thin sides. They are pans, goblets, cups, and saucers, some of which for elegance and good taste might— better than the depraved porcelain of the present century—

Fig. 9.

serve as a model to our potters. The bowl of other vases is set round with a series of five or six pointed excrescences, one of which is perforated for the cord for hanging up the vase. One of the unbroken vases belonging to this order (fig. 9) comes from Castellazzo di Fontanellato; the other, the larger of the entire ones (145 m. high, greatest diameter of bowl, 180), fig. 10, was dug up at Enzola. A small vase (73 m. high, greatest

Fig. 10.

horizontal diameter 70), with sides not so thin, and likewise adorned with pointed protuberances, but with none of them perforated, is represented below (fig. 11). Its form is that of a double hexagonal pyramid, elongated and truncated at both ends, of which one forms the base, the other the aperture, which is roughly pentagonal, and so not in accordance with the six faces which constitute the upper half of the sides. The excrescences

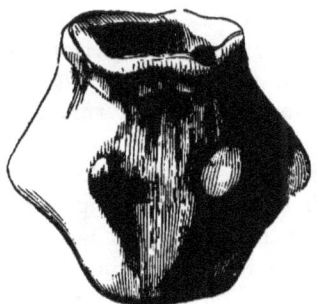

Fig. 11.

rise from the six solid angles of the bowl of the vase. Its use is unknown; probably it was a scent bottle, and perhaps had to do with burial. But the greater part of the earthenware we are discussing has not so remarkable a form, having neither protuberances nor rims, but presents the common conico-truncated forms, or else globular or hemispherical.

The edges of these earthen vessels offer modifications which we shall attempt to describe to the best of our power. We have evasive edges, or rather some more some less inclined or bent inwards and outwards: these we shall call lips: these internally are sometimes convex, sometimes plane; the plane ones again sometimes have a horizontal projection inside, and sometimes are plane; sometimes beautified with designs, engraved or raised (*in relievo*), always, however, rectilinear. Sometimes the rim is formed by a small string-course (*cordoncino*); but for the most part the sides are the same all through, which, being cut, form the margin, which is the most simple of the edges.

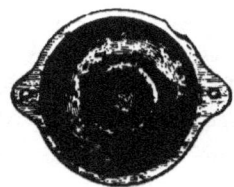

Fig. 12. Mercurago. (Internal part of Cover of a Vase.)

Sometimes this margin bears consecutive impressions, carefully made by the fingers of the potter,—something like the lacework on certain kinds of piecrust. Finally, such vases have inside, a little beneath and running along the edge, a string-course.

For the most part, the sides of the pottery being drawn in form the *base* or bottom, which in some cases is plane, in others is more or less convex. The convex bottom sometimes presents outside a little rising in the middle, either simply convex, or else with an obtuse point, which brings to one's mind the Roman amphoræ. Another similar one is, on the contrary, concave in the centre. In one of these cases in the bottom, on the inside, a convexity corresponds to a concavity on the other. A true base or support has been met with only in a few, and those, little vases; among those which have been represented, we do not see a single example. This support is only

Fig. 13. Sesto Calende.

formed by a simple little band, which, being placed round the centre of the bottom of the vase, supports it, and prevents its coming in contact with the plane on which it stands, like our common little plates (*scodelle*).

But however, rough, as far as has appeared up to this point, were the makers of this porcelain, there was in them, notwithstanding, an inclination to vary their handiwork; this especially shows itself in the different forms and different embellishments of the handles, as being the parts of the vases, which, from their thickness, escape most easily being broken into fragments The most simple form of handle would be that formed by a plain horizontal or vertical projection sticking out from the external wall. A semi-cylindrical protuberance, perforated lengthways, constituted another sort of handle, if indeed it merits that name, since it could only serve to hang up the vase with a piece of string, and in no wise to hold it in the hand. This perforated protuberance sometimes showed its greater axis and the corresponding hole horizontal, and springs, it may be, from the rim of the vase so as to form part of it, it may be below it.* Sometimes it springs from the bowl and presents the greater axis, and the little hole straight up and down. Less rough are the handles of the larger and more common earthenware vessels, used even to the present day, which, either curved, cylindrical, or else plain, are applied to the vase either with both extremities to the external wall, or with one to this, and the other attached to the rim. One variety of this type consists of those handles which, similar to those described, run horizontally along the bowl of the vase instead of from above to below, as in the preceding

Fig. 14.

instances. Of the kinds of handles hitherto noticed, few in truth display ability or good taste in the artificer. But we cannot say the same of another type, which we may call appended (*appendiculati*) handles. These, though in other respects like the common handles, bear on the upper part an appendage which sometimes supports a transverse bar, at other times grows into a half-moon shape.

The greatest development of the moon-shaped variety of this type is observable in those handles in which the two

---

* *Vide* fig. 9, *supra*.

branches of the half-moon are so prolonged as to resemble the two branches of a stag's horn smoothed down (fig. 14), such as are also found in the marl beds; so that on one of the branches of these horned appendages appearing in the excavation, it might at first sight be easily mistaken for a horn. The appended handles vary considerably in the form and length of the appendage and its dilatation. Sometimes they resemble ears (*orrechiute*), like that represented below. (Fig. 15.) Sometimes they are lance-shaped

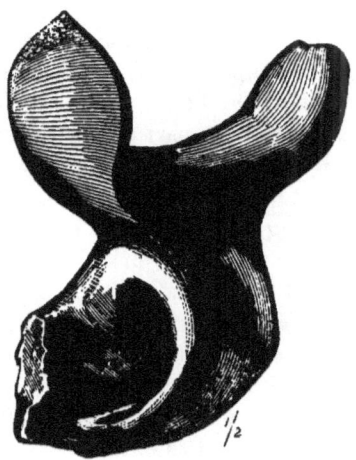

Fig. 15.

(*lanceolate.*) The design upon them, either made by scratches or string, or by pressure, is always rectilinear. Out of the large number discovered, the varieties of appended handles may be reduced to a dozen series. Of the vases dug out whole, there are not wanting those entirely without any handle, or any projection which may serve for one; they are small, even very small, so as scarcely to deserve the name of vases. In the side of a similar one, and immediately under its rim, have been pierced one or two holes, a simple and expeditious way of enabling it to be suspended by a small cord.

Of the decorations of the pottery there is little to be said. We may sometimes observe right lines scraped upon them; sometimes we may find the impressions of the finger, on which we have remarked already, disposed with a certain order and design; and others which, made with some instrument while the paste was still tender, imitate a bough with alternate leaves. Finally, the ornaments consist occasionally of string-courses more or less thick and raised, and in these we first discover attempts at curved lines.

One single cover (*coperchio*) have we been able to obtain in the terra mare; it is entire, plain, circular, with a simple common handle in the middle; we found it at Vico-fertile.

We shall end this chapter on Earthen Vessels by noticing the fact, inasmuch as it may be an interesting one, of our having excavated from the upper part of some marlbeds, pieces of earthenware, which, while they are composed of the characteristic paste of the pottery of which we have been talking, bear evident traces of having been executed on the lathe, while the others were both composed and worked with the hands. We perceive in all parts, but especially in the external walls, most regular markings (*solchi*) in depth a millimeter or more, and two or three millimeters apart. The colour of the paste is more uniform than in the preceding vases; sometimes dark grey, sometimes brown, sometimes a dark brick red; from the compactness of the paste and the uniformity of the colour, we might suppose that they had been baked in an oven. From various shards of this kind of earthenware, it appears that the vases of which they formed part had the brim inclining inwards, which we do not observe in the other Celtic vases of the marl-beds, with the vases worked on the wheel we constantly find pieces of vessels in *Pietra ollare* (pot-stone),*

---

* The fact is singular and deserves mention, that in a district of our highlands the way and secret of making a kind of pottery is jealously preserved; which reminds one in paste and baking of that of the Celts. This little district is called *Càsola* di Ravarano, and in our country is specially known as Casola of the pots (*delle olle*), because it is this place which provides it with this kind of vase. It is situated on the left side of the post and military road which leads from Parma to Pontremoli. The earth of

like saucepans; these naturally were worked upon the wheel; they are pierced at the bottom, with a view to being stopped up with metal plugs. The bottom inside, and sometimes outside also, is covered with a varnish of a black tint, of a carbonaceous nature, which proves that they were exposed to the action of fire; while the holes in the middle attest that the vase did not serve to cook any substance, at least not liquid.

---

this pot stuff contains, as far as is known, besides clay, *grains of silex*, chalk, oxide of iron, and calcareous spar, carefully proportioned, cut up into minute rhomboidal pieces, imperfectly calcined and mixed with the paste, with the object of preventing its breaking into chinks when it is exposed to the fire. The potters of Casola give to this calcareous substance the technical name of *tarso*. Many Celtic potsherds of the marlpits, also, contain particles of calcareous spar, which at first sight one is tempted to take for grains of white quartz. They do not, however, contain it in the same degree as the pottery under discussion. These, roughly composed in the way we have mentioned, are next formed with the *hand*, and afterwards baked at a common fire, never in an oven, less badly, however, than the Celtic vases, owing to which the tint of the superficies of the fracture in them is more uniform, almost the same both inside and out. These pots come out a *little* porous, and slightly refracted (*leggermente refrattarie*), and they can be exposed to an intense heat (300° about of the pirometer), and may be fifty times exposed to this heat; heated, however, till they become cherry-colour, they break up. These properties of the vases of Casola have decided in a great measure their uses: their size and form varies according to the special use for which they are intended. The *ollas*, properly so-called, have a truncated conical paraboloid form, with little ears, or lateral semi-cylindrical protuberances, perforated to admit the passage of small cords, or iron wires, with which to hold them in the hand, or hang them up,—almost exactly as may be seen in some kinds of Celtic vases. They serve principally in the country to hold the milk while coagulating; in town, in houses and shops, as a sort of small oven, in which they place live coals, either to warm themselves, or to make instruments, or other objects, red hot. In chemists' laboratories they are employed in chemical operations which need great heat; *e.g.*, the calcining of magnesia: the kind called *testi*, again, are formed in the shape of a truncated cone; rather shallow, and with a sufficiently wide base, with four lateral ears so as to hold them by means of four irons when they are burning hot; and they serve as a reverberating furnace to put over bread and cakes in cooking, they are made the breadth of a meter or more. Finally, there are made little baking vessels, suitable for the laboratory of the chemist. We owe the major part of these remarks to Sig. Berté, engineer, who, living in Ravarano, was able, from his acquaintance with the pottery of Casola, to furnish a more genuine account than anyone else.

III. Utensils.

These consist of stone, bronze, and stag's horn. We shall begin the review with the utensils in

BONE,

as those which might be confounded with the remains of various ages. They come principally from the marl beds of

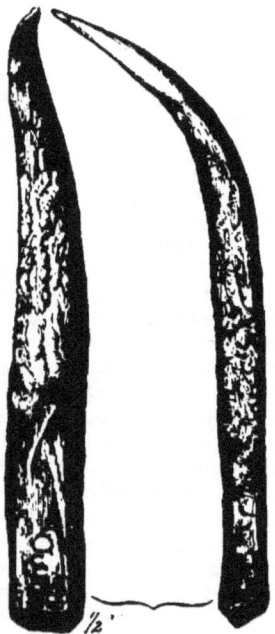

Fig. 16.

Campeggine and Castione, which, up to the present time, have yielded the largest contingent of Celtic objects;* part of them are made out of stag's horn, whether of the roe, deer, or of the stag—the common one, so called; and for this reason one meets in the marl-beds not unfrequently

---

* And for this we are indebted to the goodness of the brothers Signori Giacomo and Carlo Cocconi, of Campeggine, and the Cavalieri Ugolotti Manerini and Romualdo Ugolotti, of Castione.

such horns, with traces of the instruments upon them, which were used to cut them and reduce them to a round form. We may here observe the imperfection of the instruments employed to this end when one compares the inequalities of the surface, and the considerable number of cuts employed, with the plainness and neatness of the few cuts which one finds on the stag's horn employed for similar purposes in later times. As I have already remarked, in speaking of the handles of the pottery, we not seldom find branches of horn in " process of smoothing, and some very smooth and shining,—perhaps intended to smooth other objects, or only to bend them.

Certain pieces of stag's horn have a resemblance to coarse saddle needles (*aghi*), fig. 16, slightly bent at the end, split lengthways, and partially smoothed, but not polished, which indicate an age of roughness, and were perhaps used to interlace and adjust the nets. Of bone, also, are certain pegs (*cavicchii*), which resemble the pins of a violin, pierced in the head, which has the form of a circular disc, as well as an elliptical disc,—like that just dug up from the marl-bed of Vico-fertile (*vide* Plates), and found by Sig. P. Corbellini, who wished, very politely, to make us a present of it. Of the same substance as these pegs was found a small comb, in form like those which our peasants hang now to the walls of their rooms,

Fig. 17.

almost square (35 millimeters long to 30 wide, exclusive of the handles). One of the longer sides bears the teeth, and on the opposite side is the handle, formed of a bandlet cut out of the same piece of horn, crossing from one corner to the other of this longer side of the comb, in a semicircle. It is prettily ornamented, though simply, rectilinear designs being engraved upon it, and was excavated from the marl-bed of Fodico di Poviglio, on the property of Sig. Cavaliere Pietro Martini, to whose courtesy we are indebted. For finish of workmanship among articles in bone, the two little wheels (*ruotelle*), fig. 18, occupy the first place, in their diameter 46 to 48 millimeters, which were found in the marl-bed of Campeggine, the

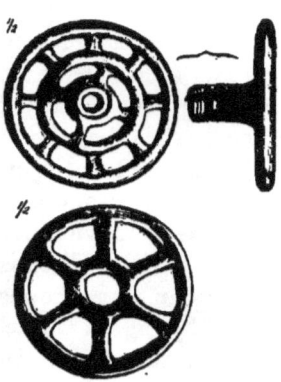

Fig. 18.

one formed of six radii and a single circle; the other, the smaller, of four radii between the nave and the first circle, and of eight between that and the second. In the smaller one, the nave projects on one side to 19 millimeters. Isolated as are these little wheels, it is impossible to divine their use. In accuracy of workmanship come next to these the handles of awls and other instruments in bronze, with engraved designs, circular and rectilinear (*vide* Plates).

UTENSILS IN STONE.

At present the objects in stone, which have been excavated from our marl-beds, are few and little varied. A single cuneiform hatchet (*accetta*) of jasper (*quarzo lidio*), or rather, touchstone (*pietra del paragone*), found in the marl-bed of Campeggine. Small, and with a nearly straight edge, we may suppose that, attached to a handle, it may have served as a chisel (*scarpello*). The other cuneiform hatchets and the axes of stone collected in Emilia, which I have seen, were either not extracted from the marl-beds, or else in some cases their precise derivation was unknown; they bear the vulgar name of lightning stones (*pietra del fulmine*), under the belief that they are the solid point of that electric stream.

The pieces of millstone (*macine*) are less uncommon; a few were composed of talco-schist and garnet-bearing micaceous schist; the greater part, however, are of the rocks of our hills,—that is of macigno (portland-stone) of breccia, or else serpentine. Some of these pieces seem to have been part of large slabs, and perhaps instead of under millstones, are fragments of hearthstones. We have had only one of those spheroidal stones channeled out in a circle, which seemed to have been used for slings, or else as weights either for nets or else in weaving. This stone comes from the marl-bed of Campeggine.

UTENSILS IN BRONZE.

These are wires (*fili*), pins (*spilli*), awls (*lesine*), chisels (*scarpelli*), and knives (*falci*). We have not had the good fortune to find any fish-hooks (*ami*) in the marl-beds we have visited; though in your Memoir we are assured that such fishing implements are found in other of these earths not examined by us.

IV. ARMS.

In their enumeration we shall preserve the same order observed for the utensils, and shall here begin the review with the arms in

BONE.

The sharpened points of this substance appear to be the heads of darts or javelins. From the kind of inequality of the surface left from the cuts, the imperfection of the carving implement is apparent, as well as the unskilfulness of the artificer; the point of one especially is so rough that, not wishing to wrong the maker, we shall be obliged to consider it as an unfinished work. We may take notice of some arms in

STONE.

But of these there is none of which we can positively say that it was disinterred from the marl earth. The Royal Local Museum of Antiquities, those of natural history in the Universities of Parma and Bologna, and the cabinet of natural history of the Lyceum of Piacenza, possess more than two dozen axes in ophite (green porphyry); lydian quartz (jasper); nephrite or kindred rock, sometimes like in form, volume, and thickness, to that figured below, sometimes larger

Fig. 19.

towards the cutting side, which is less curved, or else they are of a thickness proportionately greater. In the museum of Piacenza, there is a pick-shaped hammer (*mazuolo*) of jade (*giada*), like one found in the Imolese; and a second, apparently of a kind of granite, makes part of the geological cabinet of Bologna. In the Memoir often cited, mention is made of flint arrowheads found in the marl-beds; we have not been able to discover any. There is one, also, in the Royal Museum of Antiquities at Parma. The greater number of arms found in the *terra mare* (marl earth) are in—

BRONZE.

The most singular arm is that kind of axe or hatchet without any hole, and, in place of it, furnished with a lateral longitudinal replication, which is known to archæologists under the name of *celt*. Our axes of this kind belong to every variety of those denominated specially *paalstab, couteau-hache*. According to Penguilly l'Haridon,* the *celts* are the characteristic of the bronze age. They appear anteriorly to the short swords (*daghe*), to the lance-points and javelins, and appear not to have served as arms, but as cutting and cleaving implements, responding to all the wants of the people come from the east, and who formerly inhabited Europe. In the Local Museum of Antiquities, there are ten of these axes; but it is to be deplored that, for the greater part, no record has been kept of their place of discovery; two came from the marl-bed of Noceto, and one from Collecchio. Below is the figure of one of the first. The *paalstab*, excavated

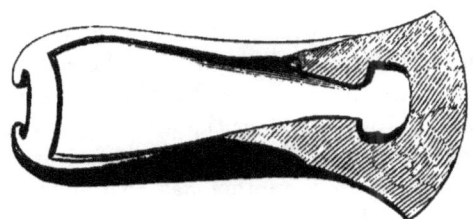

Fig. 20.

from the marl earth of Castellazzo (fig. 21), which forms part of the Giordani collection, has the cutting part much curved and large; the lateral replications, on the other hand, are short and elliptical. Over and above all the other blades of daggers and javelins, as you may call them, we have had one from the

---

* Note sur l'emmanchement des haches de bronze.—*Revue Archéologique*, Nov. 1861, p. 329.

marl-bed of Castellazzo, presented by Sig. Conte Luigi Sanvitale. It has the form and length (134 mill.) of one you

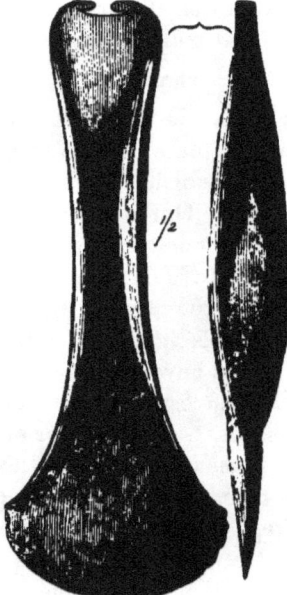

Fig. 21.

have represented, but differs from it in the elongated elliptical rather than circular hole, intended for the passage of the nail used to fasten the blade to the handle, in which it was mortised. In the Museum of Antiquities, eight similar blades of bronze are preserved, all of uncertain derivation; one has a single hole for the nail, another has two successive ones, disposed longitudinally, sometimes both on the same horizontal line (fig. 22). One of the blades, with two longitudinal holes, offers

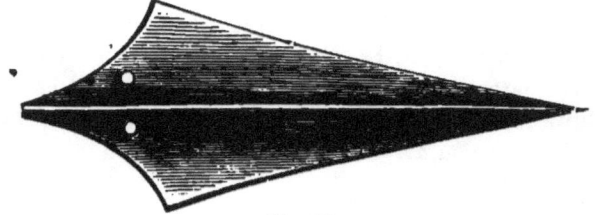

Fig. 22.

in the middle a rib or longitudinal keel, is much longer than the others, and so approaches in resemblance the head of a lance. One head, which is unquestionably that of a lance, was found some time since, and made prize of by one of us, in the marl-bed of Bargone di Salso, and we give its figure from

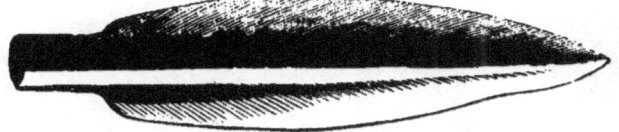

Fig. 23.

the collection of the Museum of Antiquities; its length is 16 centimeters, breadth 3·3 cent. at its greatest diameter; it seems identical in form with those disinterred at Alise Sainte Reine, which have been preserved from the Gallo-Roman epoch.* Two only of the arms in

IRON,

up to the present time, collected from the marl earth, appear to us to be anterior to the Roman epoch, and are—an arrowhead of octahedronal shape (*ottaedrica*), with a transverse section of a crushed rhomboidal shape (*a sezione transversale rombica schiacciata*), and a sort of small knife at the convex back, with a short appendage to insert in a handle; the length of the blade reaches 19·2 centimeters, and the greatest breadth is three centimeters. These arms come from Castellazzo di Fontanellato.

In nearly all the marl-beds are found scoriæ (*scorie*), sometimes flinty, sometimes ferruginous,† not alone of stones which show the sudden action of fire, which proves that the people to which we owe these azotised (*azotate*) deposits knew the art

---

* *Revue Archéologique.*

† Professor Galeazzo Truffi took upon himself very kindly the labour of analysing some of these scoriæ; he found no *copper* whatever, but found *iron*, establishing that they were composed of silica and iron in certain proportions (*di silicato ferroso-ferrico*), so that he considered that they were the scoriæ of mines or refining houses, or similar works.

of founding in metals. And this is further put beyond doubt by the fact that moulds (fig. 24) are found in steatite (a rock of our Apennine) for the making, by fusion, of the arms in bronze, as well as unformed pieces of this alloy.

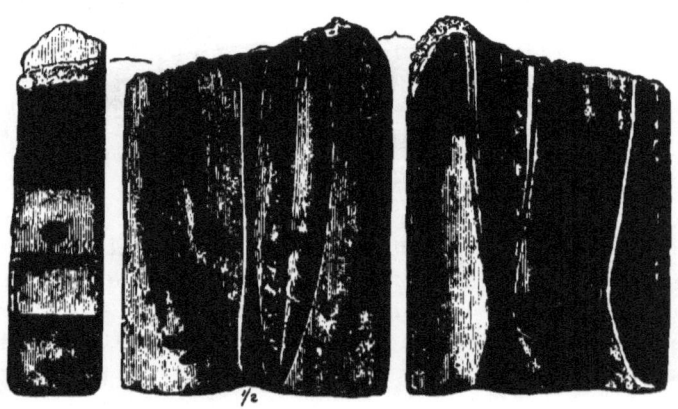

Fig. 24.

V. OBJECTS OF UNCERTAIN USE.

SPINDLE-WHIRLS (*Fusajuole*).

There are very many of these objects, for the greater part of terra cotta, more or less discoidal, or conical, or spheroidal, pierced in the centre, to which the archæologists of France and Germany, as well as our own, have given the name of spindle-whirls (*fusajuole*). The paste of the spindle-whirls is not, for the most part, equal to that of the earthenware; instead of the grains of sand, we find powdered carbon and ashes; the colour is ashy in the internal parts, and ash-colour varying into yellow and red on the outside. Some few spindle-whirls are black, and of a substance probably similar to the thinner vases, and, like a great number of these, are shining externally as if with varnish. They are very various in form; and although eight different ones have been represented by you, from those which, in the course of the summer, we sent from Campeggine, courteously presented by the brothers Cocconi, not one represents the other six, collected, in the sequel, in the marl-beds. Some few bear ornamental

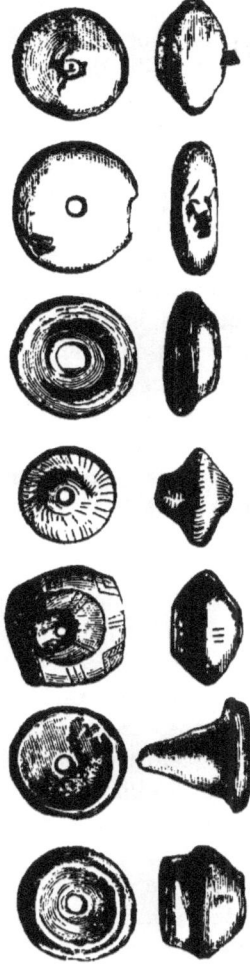

Fig. 25.

marks scratched upon them, and are among those you have had engraved (fig. 25).

Besides all the spindle-whirls of earth, there were dug up

from the marl-beds of Castellazzo di Fontanellato three others, which are cut out of different substances. One was made out of a stag's horn, it is in the shape of a cone, and is very highly polished; the second of steatite, of a greenish tint, and spheroidal; the third, of a whitish limestone (*calcare*), is disc-shaped, brought to a high degree of polish, and certainly manifests an advanced epoch in art among the people who used such implements. Among the objects in the Museum of Antiquities at Parma, which are of uncertain derivation, there are twenty spindle-whirls, some in limestone, steatite, and even *amber*, but the greater part are of earth; some are polished, some are ornamented with circles, concentric with the hole pierced in them, or in concentric lines disposed in groups on the back of the spindle-whirl. We find among these the transition from the more depressed discoidal form, almost medallion (*nummulite*), to the acute conical. Some one of those in terra cotta *is said* to have been collected from the ruins of the Roman city of Velleia. The different forms, finish, and substances of the spindle-whirls would lead us to suppose that they must have served for various uses in proportion to their diversity; perhaps the most beautiful and carefully worked were amulets, or else buttons; the others, weights, used either for nets or in weaving.

It is customary to separate from the spindle-whirls those cakes or discs of terra cotta pierced in the centre, and composed of the same substance as the greater number of the former, but which are considerably larger and heavier; their greatest diameter is 15 centimeters, their greatest thickness 56 millimeters, weight 2·230 kilograms.* It may be, perhaps, that these articles served for different uses, for example, as weights; for we have found one in the marl-bed of Fodico di Poviglio in which flutings are discernible, which separate like rays from the hole in the centre, and which appear to be the impressions left by the friction of rope. The fact of having dug up one from the bottom (*punte*) of the piles of the lake

---

* The *kilogramme* = 32¼ oz. troy, or 2·2057 lbs.

habitation of Castione, would lead one to suppose that they served to keep nets steady. Some of these objects are plain above and below, that is discoidal; others are convex above, like buns (*focaccie*). It is well known that pieces of charcoal are found embedded in the paste, and in one is found embedded even a snail,—the *Cyclostoma elegans*,—which shows how careless were the makers of such potteryware.

Besides all the earthenware and all the spindle-whirls which we have spoken of, we meet in the marl-beds with other small objects in earth, badly baked, in form sometimes *disc*-shaped, *without* any *hole*; sometimes ball shaped (*pallottola*), of which it is impossible to divine the use which they served.

Organic Remains.—Man.

In the foregoing report on the lake habitation of Castione (p. 20), it was remarked, in speaking of those layers in the bed of marl-earth, that there were in past years discovered two human skeletons, lying on their backs; their bones, as is generally the case, have been dispersed. However, fortune in the sequel has decided that one should come into my hands; it is a right radius (*radio*), 243 millimeters in length. From this measurement we might argue that the person to whom it belonged reached 1·67 meters in height, or thereabouts. From various other marl-beds, however, we have human bones, but excavated from the superficial or upper strata, and mixed with objects of Roman industry or of a more recent epoch; so that they do not offer the appearance of fossilization, to which the radius in question has attained, which is comparatively heavy (58·3 grammes*), hard, and ashy in colour. The discovery of this human bone would lead us to consider as true the assertion which was hazarded regarding the beds at Castione, as well as in the particulars referred to above, of the probability that, by continuing the excavations in that *marl-bed*, we might discover other skeletons of the inhabitants of the

---

* The *gramme* (unit of weight), 15·44 grains.

palisade; and therefore, on this account, over and above that of the palisade itself, it will be in the interest of science and sufficiently important, that the excavations at Castione should be undertaken, not as up to the present time, only with an economic and agricultural view, but on scientific principles, and under the superintendence of a person of intelligence; for this naturally is also wanting no little pecuniary means, which, as you know, is too often wanting.

Mammalia.

The bones which are met with in the marl earth, like those of the lake habitations of Switzerland, are little fossilised, or rather, have lost a very small portion of their organic substance. We find them generally broken, even reduced to small fragments; those only of the extremities are, for the most part, entire. We have not been able to find a single skull complete, nor yet the isolated bones of the cranium, excepting the *pieces* of a frontal bone (*frontale*) and of an upper maxillary (*mascellare superiore*). Such a circumstance sufficiently increases the difficulty of determining these remains, and the more, that for the greater part of them we are not concerned now to establish to what species, but to what variety or race of a given species they belong; and this difficulty is increased much more by another fact which has been observed in the *marl-beds,* as well as the pile habitations,—namely, that of certain species there are no bones discovered which are not worked or converted into instruments, as they were preferred to softer bones of other kinds: such worked bones are, of course, unserviceable for the *relative* measurements which are required for the determination of race.

From the facts mentioned we may deduce,—

1st, That whatever the agency may have been which heaped these bones together in the marl earth, they represent *sometimes* the remnants of the *feast,* whether sacred or profane, for which the animals were used, to which the bones appertained. But as from these remnants of the feast men withdrew for the most part all that could serve for the objects of their arts; so *sometimes* these heaps of bone are the collection of the refuse (*scarti*) of industrial works, or rather of the remains of the same. We see, then, why the greatest part of the bones are not entire or intact,* and this is why we find the bones of all sorts of animals, domestic and wild, mixed together confusedly.

Secondly. That the populations for whom these mammalia served either as food, or to supply their workmen, or as their helpers in the chase, broke the crania to extract the brains. What is more, as far as we have observed, they never forgot to open the alveolar cavity of the inferior maxillary, especially in the dog, violently wrenching off the vertical branch of the mandible from the horizontal to get at the pulpy substance.

Finally. They no less spared the cylindrical bones, since they broke these up to extract the marrow. Analogous facts have been already observed, as you have mentioned in your Memoir, by Morlot and others, in the bones of the lacustrine habitations; and it also from this appears that the inhabitants used the brain and marrow either gastronomically or technically,† as, at the present day, certain savage people of North America.

The Swiss naturalists were more fortunate, as far as appears, than we have been, in so far as in the magazines of bones of their pile villages, they found not rarely crania, even though deprived of the facial bones; whence it would seem that the people of our marl-beds, in the operation of scooping out the brains of the animals, proceeded in a different way from that in use among their Helvetic brethren, although they agreed with them in tastes and general customs.

Domestic Mammalia.

Among the remains of the most ancient lacustrine habitations of Switzerland, as of the other deposits of the so-called

---

* This is true of the scapulæ, the plates (*laminæ*) of which must have served for the fabrication of utensils of large and plain superficies.

† To dress skins.

age of stone, the number of the bones of *wild* animals by far exceeds that of the bones of *domestic ones;* but by degrees, as we come to the strata of less remote epochs, the quantity of bones of savage mammalia diminishes, until, when we arrive at the deposits of *our age*, the bones of the domesticated animals decidedly preponderate. From this we should infer that in the first ages the mammiferous savage fauna considerably exceeded the domestic; and in the later epochs, the first diminished by little and little until, in the historic times, they were obliged to cede the field to the domestic fauna. The scarcity of the bones of the suckling domestic animals in the remains of the most remote ages, and the consequent conjecture as to the limited number of such animals then living, will receive obvious explanation in the sequel. In the first beginnings of peoples and of their civilisation, the species can be but very few, and still fewer the races, of mammifers subdued and then domesticated; and their education would have to strive with many obstacles both on the part of nature, the instinct of the animals, and of external agents. Wishing, besides, to make use of them for various wants, domestic, industrial, and agricultural, seeing that the wild mammalia were not adapted for this, they came to use these as food, and to spare the others, even in advanced age.

On the other hand, it is well known that the scarcity of the bones of wild animals, at least of certain species, in the deposits of the less remote epochs, as well as in the marl earth, may depend upon another circumstance already noted by us. The bones of some animals of this kind, as of the wild boar and the stag, being stronger and more keen at their fractures than the others, and notably more so than those of all kinds of domestic animals, came to be employed in preference to them in utensils and arms. And for this reason,—although as we have just said, the bones and the other remains of wild mammals are scarce in the marl earths, while without comparison the remains of domestic mammals are found in abundance,—we cannot from that *sole* fact allow ourselves to be led to hold that these deposits have had their origin in historic times. Some mammifers, as well wild as domesticated, seem to have come only *accidentally* into the hands of the inhabitants of our *marl-beds;* for example, the bear, and perhaps also the horse. In Switzerland, the bones of this solid-hoofed beast are sufficiently rare among the remains of the palisades, and especially of the more ancient,* and we cannot with truth say they are common with us.

We shall begin the review of the few species of domestic mammalia with

The Dog (*Canis*).

This faithful companion of man in every stage of civilisation and in every climate, although usually it only serves as an animal for the chase, as a watch, and for amusement, and does not appear to have been otherwise employed by the inhabitants of the marl-beds; still, from the absence of crania, and from seeing their few cylindrical bones broken, we are induced to suspect that their brains and marrow were applied to the same domestic uses as those of other animals. Perhaps when the dog was become old and useless they killed him, as is now the custom among the Esquimaux; or perhaps, like the Chinese, they ate him after he had died; however that may be, the bones of the dog are rare in our excavations, as well as in those of Switzerland. In the last mentioned country, they have had the advantage over us of finding skulls, in a good state of preservation, of this carnivorous animal; also, in number proportionately greater than of any other mammal, and belonging to adult individuals, and even old;† so that

---

* According to *Rütimeyer*, the remains of the more common species of mammals present themselves in the excavations of the lake habitations of Switzerland in the following quantitive distribution :—In the most ancient and smallest palisades, follow one another in decreasing series, *the stag and wild boar, the ox, the goat, and the sheep.* In the less ancient and more extensive, instead, we find the ox, pig, stag, sheep, and goats. In the marl-beds this last scale in the number of the said species of mammals is almost exactly the same.

† By which the opinion hazarded above, that the dog did not belong to the category of alimentary animals, will be confirmed, since these are eaten young.

M. Rütimeyer was able to compare them with the heads of living dogs, establish their analogies, and determine their race.*

The canine remains of the stone period belong *exclusively* to one dog, which, in the form of its cranium and of its face, stood between our hunting dog (*bracco*) and the setter (*cane da ferma*),—the author calls it the dog of the turbaries (*canis familiaris palustris,—torfhund*), and suspects that it was the type of the species. The canine bones of the later ages belong, on the other hand, to a much taller race, and akin to the butchers' dog (*cane da beccajo*). From the excavations of our marl-beds we have only three inferior jawbones, and the lower piece of the shoulder. Comparing the measurements of the first with those given by Rütimeyer of his turbary dog,† we come to the conclusion that these remains belong, if not to that same race, at least to one very nearly allied to it,—certainly not to the larger kind mentioned above; and there seems to have been only one race of the marl earth, as the three bones are exactly the same.‡

Pig (*Sus*).

Of this *species* we have found abundant spoils,—teeth, pieces of jawbones, of scapulæ, humeri, ulnæ, radii, metacarpal bones, tibiæ, and astragali, in part bones with epiphyses, (cartilaginous protuberances contiguous to the bone), and milk-

---

* *Fauna der Pfahlbauten in der Schweiz*, Basel, 1861, in quarto (with woodcuts), and six tables lithographed.

† Length of the jawbone from the angle to the margin of the incisor teeth,—

    Marl-bed of Bargone.................................... mill. 117
       ,,   Casaroldo .....:......................... ,, 125
       ,,   Fodico .... ,, *anterior part wanting.*
    Pile habitations of Switzerland 110–120.

‡ There is only a little difference in the distance between the first premolar teeth and the canine, which is 4 mill. in the one from Bargone, and 7 mill. in that from Casaroldo.

teeth; of the inferior mandible we meet with the horizontal branch, with the beginning only of the vertical,—and be it understood, with the sockets of the teeth opened inferiorly (*inferiormente scoperte*). We hold that almost the whole of these remains point to *domesticated* individuals; for the bones have a fatty brightness as well as vitreous, and their surface is rather polished than rough; the talon of the third molar tooth is much retracted, almost wasted away, which are exactly the characters held to indicate a state of domesticity in the species.

Now, it would be advantageous to settle whether to either of the two races, met with in the lacustrine habitations in Switzerland, we can refer the remains of the pig of our marl-beds. One of these is the common pig (*majale—sus scrofa domesticus*), to which is given, as stock, the wild boar (*cignale—sus scrofa ferus*), a race known to all; the other was named by Rütimeyer the pig of the turbaries (*torfschwein—sus scrofa palustris*). The wild stock of this race, common in the age of stone, no longer exists; it has been extinct from the earliest historic epoch. One meets again with *traces* of a similar race, but in a domestic condition, in some parts of Switzerland, that is, in the mountains of the Grisons (Grigioni,—*Oberland**); and this domestic pig sufficiently small, with short legs, ears not hanging and small, snout short and thick, bristles long and rare, of a uniform and blackish colour or dull reddish brown. The *Siamese*, or *Indian* pig, approaches very nearly to that of the Grisons and that of the turbaries; but of this, too, there are only known those in a state of domestication. All the three races mentioned, or perhaps modifications of a single race, are distinguished from the common pig by the following osteological characteristics:— the molar teeth more robust, and the premolars more weak,

---

* "Nei Grigioni (*Oberland*)." By this, apparently, Messrs. Strobel and Pigorini intend a particular part of the Grisons,—probably the line of snow mountains skirting the Engadine on the west, though I am unable to find that term applied to it.—EDITOR.

the hinder molars and premolars approximated (*ravvicinati*), and the dental system, consequently, more compact,—the incisor and canine part of it reduced by the shortening of the premaxillary bone, and of the symphysis of the chin. The pig of the turbaries will represent the wild form of the animal, the other two the domesticated form, which, thanks to cultivation, exceeds the other in volume.* It would seem that the pig of the turbaries was rather herbivorous, while the common pig (*majale*) is decidedly omnivorous, which one might argue from the comparison of the superficies of their respective maxillaries. It seems further, that, judging from the size of the orbits for their eyes, the first was a nocturnal rather than a diurnal animal, at least in a state of independence: the wild boar is crepuscular. The pig of the turbaries was brought under the rule of man towards the end of the stone period, before we meet with signs of the domestication of the wild boar (*cignale*), although both one and the other, as free animals, appear contemporaneously on the scene of the pile dwellings of Switzerland.

Another pig, nearly related to that of the turbaries, and perhaps nothing but a rather smaller variety of it, appears, as would seem, in the age of bronze, and lasted certainly to the time of the Romans, and to the sixth century of our æra, and so must have existed for some time with the domestic *typical* pig of the turbaries. Some people hold it to be the original stock of the Berkshire pig. Only from the porcine remains of the *marl-beds* of Bargone di Salso, of Basilica nuova, and of Collecchio, and from those dug from ground of the same nature in the city of Parma, have we been able to discern our pig (*majale*); rarely, also, from the spoils of the pile habitations in Switzerland. The bones and teeth of pigs from all the other *marl-beds* point to the pig of the turbaries, as, with regard to the teeth and jaws, you may judge from the following

---

* Rütimeyer, *l. c.*, p. 189.

TABLE OF RELATIVE MEASUREMENTS.

PARTS MEASURED. MEASURE IN MILLIMETERS.	Small varieties of Pig of the Turbaries.	Typical Pig of the Turbaries.	Pig of Marl-bed.	Pig of the Grisons.	Common Pig. Majule.
LOWER MANDIBLE.					
Series of molar teeth	112	123–128	129	122–131	133
Same without 1st premolar	92–97	102–112	106	104–110	112
Three molars together	52–66	65–74	65	71–72	72
Third molar alone	26–32	33–37	30–35	32	36
Molars 2, 1, premol. 4, 3	43–64	55–64	58	59–67	66
3 Posterior premolars, 4, 3, 2	33–39	25–40	37	37–40	40
Dist. between premol. 1st & 2nd	... ...	10–13	14 (?)	8–14	13
,, ,, 2nd and incisors	... ...	37–47	50	47–56	56
Greatest diameter of the socket of the canines	... ...	10–17	14–16	15–19	16
Dist. from the said socket to the point of symphysis	... ...	30–37	29–31	38–44	41
Dist. across the external margins of sockets of the canines	... ...	44–53	42–47	55–58	53
Length of symphysis	58	62–79	63–65	73–90	75
UPPER MAXILLARY.					
Length of three molars together	60–67	65–77	63	69–72	70
,, of 3rd molar alone	26–34	30–40	29	31–32	32
,, of molars 2 and 1, premolars 4 and 3	56–60	59–63	60	62–66	63

The relative measurements of the pig of the turbaries and its lesser variety, as well as those of the pig of the Grisons, are copied from the above-mentioned valuable work of Rütimeyer; the measurements relating to the common pig are taken from a skull in the Cabinet of Natural History of this university, in which the third molar has not yet entirely grown, the posterior part of the heel being still concealed in the socket. The measurements of the pig of the marl-beds were taken from part of a mandible of Vicofertili, and from another reconstructed from Fodico di Poviglio, in which the teeth are separately used; whence result the greater difference in this part of the skeleton between the pig of the marl-bed and the majale,—the first being adult, and the second, though larger, still young,—for the measurements of the upper maxillary served the part of one found at Castellazzo di Fontanellato, unfortunately of an individual still young; isolated maxillary teeth of both the maxillaries, obtained from the *marl-beds* of

Madregolo, Castione, and Colombarola di Fontanellato, helped to complete these measurements, which must be confessed to be imperfect. Very few incisor teeth—canine and premolar—of the upper jaw, have as yet been met with in the marl earth.

The Horse (*Equus*).

From the pile habitations of the age of stone, which in Switzerland furnish so much material for the palæontological study of the existing geological epoch, the bones of the horse, as we have remarked, are *almost* entirely wanting. In the later age of bronze, this solid-hoofed beast appears less rarely, but still accidentally; they are, perhaps, the remains of some trophies or some prey taken from the enemy; and it is natural that the horse should not be frequently found there, since the habits of those lacustrine people were certainly not propitious to their breeding. We have also noticed that in the *marl-beds* the spoils of the horse are not certainly rare, but still not very common, perhaps for the reasons given above. The cranium, as usual, is wanting; we have some inferior mandibles, isolated teeth of the upper maxillary, neck vertebræ, radii, metacarpi, metatarsi: they belong, like the horse-bones of the lake habitations, to the living species, *Equus caballus*. But as in those, so in the marl earth, we find two races,—the one smaller and with finer extremities, like those of the mule and ass, which we may call the fine race; from the state of fossilisation of its bones we might suppose it was more ancient than the other; and it is not beside the question, perhaps, to remark that the species of horse of the drift (*diluvio*) (*Equus angustidens*) had the extremities very small, and with resemblance to the ass: the other race, more commonly represented in the marl-beds, is larger, more coarse, and its bones less fossilised; it is, consequently, later than the first.*

---

* Here are the measurements of the few entire bones of both races:— *Small* Race.—Metatarsal, long, 254 mill., from Pontenuovo. *Large* Race.—Radius, long, 345 mill.; Metacarp. long, 218 mill., from Beneceto, Bar-

We have not yet met with traces of the *ass* in the *marl earth*, unless we refer to this species the *piece* of a lower mandible, as usual, with the sockets exposed, which belonged to an individual of considerable age, and which was dug up from the marl-bed of Castione.* In the Helvetian lacustrine habitations, the ass appears after the horse, and in still less numbers.

The Ox (*Bos*).

Without comparison, the greatest number of bones of the marl earth belong to this species. In Switzerland, even from the age of stone, there existed two races, in size opposed to one another,—as at the present day the little Breton cattle to the larger ones of our country,—and both lived together in the same locality, not separated by geographical limits; and this is exactly the same in our *marl-beds*. We shall begin the discussion of them with the smaller race, as being that which, as far as we know, is extinct among us;† while in the central cantons of Switzerland there are still some traces of them (in the *braunvieh*). Rütimeyer calls it the cow of the turbaries (*torfkuh*). It is the most ancient domestic animal of the pile habitations of Switzerland, since although the other race came to be domesticated also in the age of stone, it was at a later period than this. Besides the smaller height and length of its body, it was singularly distinguished by its fine and active extremities, with delicate phalanges (*falangi*), which doubtless bore sufficiently small hoofs. In the form of its mandibles and the fineness of its limbs, it came near the zebu (Indian ox), and like that to the stag. We may see that they were akin to the little Breton race of the highlands of Scotland and Wales, which, at the time of the Roman invasion, com-

---

gone, Castellazzo. Measure of the greatest upper molar tooth:—height, 90 mill.; length, 30 mill.; breadth, 28 mill.; locality uncertain.

* This branch of maxillary contained only the premolars and the first molar only; their united length 105 mill.

† The ox called *Bardigiano* of our mountains appears to have nothing in common with these, except lowness of stature.

posed the herds of those warlike people in Britain; the short-horned oxen of Finland are also allied to them. The wild stock of these kindred races, not to say varieties of the same race, appears to have died out before the historic epoch; its fossil remains are met with in the pliocene strata, modern and new, of England and Scandinavia, also in the turbaries of Ireland,—it is *Bos brachyceros*. The type of its domesticated race, the cow of the turbaries, ceased to exist, at least in Switzerland, about the sixth century of the new era. We give below the measures of the various parts of the skeleton of the small ox of the *marl-beds*, against which we have set the measurements which Rütimeyer gives of the same ox in Switzerland; the former will be the complement of the latter, as we have been able to measure some bones, of which the said author makes no mention, perhaps because he had not pieces of them which would give the measurements. We have also set the measurements, relating to the other ox, the greater one of our marl-beds, against those of our common oxen,* and also those of the great race living in Switzerland, and have added as well the few measurements which the above-named author has given of certain remains of an ox still less than that of the turbaries. These were in part taken from the border of the Lake of Constance, near Steckborn, a place of uncertain archæological date, and in part dug up from a mound (*mammellone*), which, in the sixth century after Christ, as would appear, was raised with a view to performing sacrifices, near Chavannes in the Vallais; and we have added these last measures to show that the little race of oxen of the marl earth would seem to have been the mean between that of Steckborn and Chavannes and that of the pile habitations of Switzerland. Finally, we have noted the measurements of certain bones of stags, some fossil,—of the Swiss pile dwellings,—some living, which show the analogy mentioned above between them and the *Bos brachyceros* of the *marl-beds*.

---

* For these measurements were employed two skulls; one of an ox, the other of a cow, and some of their bones, which are preserved in the collection of Natural History of the University of Parma.

Besides the parts measured of its skeleton, there have been found pieces of ribs, of pelves, of lumbar vertebræ, of bones containing epiphyses, and of the mandibles of calves. A metatarsal bone, with exostosis, proves that such a morbid state of that domestic mammal was not unknown even in those remote times. The inferior mandibles, as we have already remarked to have been the case in those of the pig and the horse, are deprived of the ascending branch, which is found separately; these are open from the side of the sockets, precisely as was observed of the cow of the turbaries in the lake habitations of Switzerland.

From a glance at the table of measurements, you may suspect that the remains of the greater race of oxen, of the ancestor, as far as we can see of our common bull, are—as they are in fact—much less frequent than those of the lesser ox. Rütimeyer gives, if we mistake not, to this, as stock, the *Bos primigenius*, common in the drift (*diluvio*), together with the elephant, rhinoceros, and hippopotamus; but from that time in the age of stone, in Switzerland, not on an equally wild footing with the above-mentioned pachyderms. Beside the bones of it, the measurements of which we give, we have found in the *marl* earth isolated maxillary teeth, carpal bones, the trochlea of a femur 97 mill. wide, and the bones of a calf. From these few remains it would appear that this race of domestic oxen was not, in size, either greater or less than that now existing.

PARTS MEASURED IN MILLIMETERS.	Small var. of the Cow of the Turbaries.	Cow of the Turbaries.	Small Ox of the Marl-Beds.	Large Ox of the Turbaries.	Our Bull.	Swiss Ox, larger race.
Bony Stem of the Horn.						
Circumference at base	90-105	120-155	140	130-182	150-194	162
Diam. of base, vertical	26-28	34-43	36	39-51	41-60	46
,, ,, horizontal	31-40	43-55	50	42-64	53-65	56
Inferior Mandible.						
Breadth of incisive mar.	... ...	... ...	68	... ...	75-77	79
Length of symphysis .	... ...	... ...	61	... ...	74-84	70
Height of jawbone:						
before 1st premolar	... ...	... ...	27-36	... ...	35-40	42
after the symphysis	... ...	... ...	22-26	... ...	30-32	29

TABLE OF MEASUREMENTS.

PARTS MEASURED IN MILLIMETERS.	Small variety of the Cow of the Turbaries.	Cow of the Turbaries.	Small Ox of the Marl-beds.	Large Ox of the Turbaries.	Bull of Italy.	Swiss Ox, large race.
Distance of first premolar from incisive margin	......	......	104	......	125-147	135
Molars, second, long	......	......	20	......	26-27	28
,, ,, broad	......	......	13	......	15-16	13
,, first, long	......	......	19	......	22-23	26
Premolars, second and third united, long	......	......	34	......	40-41	55
Atlas:—body long	......	......	35	......	......	436
wings long	......	......	140	......	......	16
				Greater Ox of the Marl-beds.		
Scapula, glenoid cavity, smallest breadth (transversal)	......	......	35	50	43	......
,, ,, greatest length (longitudinal)	......	......	50	65	61	......
Humerus:—breadth of trochlea	64	70-73	58-72	77	......	83
				Large Ox of the Turbaries.		
Ulna:—length from olecranon to the front	......	......	90 ?	......	......	125
,, upper margin	......	......	42	......	120	56
,, smallest breadth	......	......	37-38	......	55	42
,, height of the sigmoidal fossa	178	179-182	162-190	214-220	40-41	225
Metacarpus:—greatest length	......	45-50	45-49	68-69	......	70
Diameter of the upper tuberosity	......	46-53	45-50	68	......	70
,, of the trochlea	......	26-28	23-35	40-41	......	40
,, of the diaphysis						

TABLE OF MEASUREMENTS.

PARTS MEASURED IN MILLIMETERS.	Small variety of the Cow of the Turbaries.	Cow of the Turbaries.	Small Ox of the Marl-beds.	Large Ox of the Turbaries.	Bull of our country.	Swiss Ox of large race.
Tibia: greatest length	......	......	32	......	......	......
Breadth of upper tuberosity	......	87	80-83	95	......	102
„ of superficies of articulation with astragalus	......	40	36-43	......	......	47
Calcaneum: total length	105-115	124-135	108-110	......	......	166
Height of swelling (*tubero*) at the base	......	40-43	32-33	......	......	54
Astragalus: total length	60	62-65	59-60	......	......	74
Diameter of inferior surface of articulation	......	37-40	35-36	......	......	45
Metatarsus: total length	214	......	190-216	......	244-270	......
Diameter of superficies of articulation:—						
„ of upper „	......	......	38-48	......	53	......
„ of lower „	......	52	43-50	......	59	60
„ of the diaphysis	......	26	20-26	......	30-34	30
	Stag. Fossil.	Stag. Living.				
Greater Phalanges (*Falangi*): diam. of sockets (*glene*)	......	......	31	33	......	......
„ of the trochlea	......	......	30	36	......	......
length „ „	65	48	60	70	......	......
Smaller Phalanges (*Falangine*): length	47	35	34-37	45	......	......
Diameter of sockets	......	......	25-27	35	......	......
„ of trochlea	......	......	18-24	27-30	......	......
Radius: greatest diameter of upper surface of articulation	57	44	68-69	74-75 ?	77	80
Total length	325	250	258	......	300	320

The Goat (*Capra.*)

We have already remarked how, among the remains of the domestic animals of the age of stone, those of the goat preponderate over those of sheep the more we are led into a remote epoch; notwithstanding this, old legends allude to sheep and not to goats, probably because the latter, owing to its nature and habits being somewhat independent and indocile, never came under the influence and yoke of man so as to merit a position among the animals truly domestic, in the same way as the gentle sheep; the rearing of the latter is, therefore, much more attended to, and its races are therefore more varied and numerous. On the other hand, we find slight difference between the living races of goats, and very slight between the goat of the pile habitations and the marl-beds and that actually bred in our country; the differences reduce themselves to those of stature alone. Similarly to the pig and the ox proper of the marl-beds, the goat of those deposits is less than the kind now living in those districts, which will in part appear from a comparison of the following measures of some of their bones with the well-known ones of the common goat, *Capra hircus*, of which the goat of the *marl-beds* is a more pure expression,—perhaps the type :—

Bony stem of Horn: length along greater curvature	150 mill.
Lower Maxillary (of an individual more than adult) : length of series of molars	66 ,,
(The same of the goat of the Swiss lake habitations	69-72 ,,)
Humerus: breadth of trochlea	26 ,,
Radius, total length	131-146 ,,
Least breadth of diaphysis	15-16 ,,
Tibia: total length	188 ,,
Least breadth of diaphysis	13-15 ,,
Calcaneum, total length	43 ,,
Astragalus, length	25 ,,
,, breadth	18 ,,

Some pieces of bone, unserviceable for measurement, belong to vertebræ, metacarpus, ribs, and metatarsus, etc.; there are not wanting inferior mandibles and bones of kids. With regard to the mandibles, it is as well to remark, that they are all cleft asunder with the wonted skill, while in Switzerland they have been frequently found entire. With regard to skulls, we shall not speak more, as it henceforward seems utopian to hope to find any of mammalia, except small ones, in the *marl earth*; in this, as we have said, the *savants* in Switzerland have been more fortunate.

SHEEP (*Ovis*).

As soon as man discovered the way of spinning and weaving wool, the sheep supplanted, in the domestic and pastoral economy, its elder sister, the goat. It is difficult, if not impossible, to distinguish the bones of the sheep from those of the goat; the limbs of the latter are certainly more slender, the points of attachment of the muscles are more marked, the substance of the bones is drier; these are characters which indicate the wilder goat, or rather, the less domestic of the sheep,—characters, however, which are only relative. It is an easier matter to discover the difference between the teeth of one and the other, as we are taught by Rütimeyer, that when the greatest molars are found fixed together in the jaw, those of the goat are more serrated one with the other, and the internal part of one covers the next to it, like a tile. The difficulty of distinguishing the remains of sheep from those of goats, is considerably augmented in those which are dug out of the *marl earth*, as for the most part the teeth are isolated and the bone in fragments, as is also the case in the pile habitations of Switzerland. There, however, the inferior mandibles are frequently entire; nor are skulls uncommon though they are not complete, exactly as we have had occasion to remark regarding the bones of goats. And the difficulty in question increases still more owing to the circumstance that the sheep of the *marl-beds*, like those of the pile habitations of Switzerland, was small, with extremities fine, active, and long, by which it approaches very nearly, in stature and construction of body, to its contemporaneous goat of the *marl-beds*; but more, it had also horns like those of the goats. We find these characters at the present day in the sheep of that part of the Grisons (*Grigioni*), called the *Oberland*, where, as we have seen, are preserved traces of the pig of the neighbouring Celtic pile habitations; also, in the Shetland islands and in the Orkneys exists a small race, with thin legs, short tail, and goats' horns. Among the Welsh mountains may be met with small half-wild sheep, with horns like goats. Sheep with goats' horns are also met with in the Island of Cyprus; but among us there is no longer a breed which resembles that described of the *marl-beds* and pile habitations, which Rütimeyer calls the sheep of the turbaries (*torfschaf*), (*Ovis aries palustris*), and which he suspects may be the type of the species.

In the diluvial caverns of Alais, in the south of France, has been found the stem of a sheep's horn, which resembles those in discussion, and belongs to the *Ovis primæva*, and perhaps this is the savage type of the other. From the animal remains of the *marl-beds*, we may refer to the small sheep of the turbaries some upper maxillary teeth, a piece of frontal bone with the upper part of the orbit, and two bony stems of horn.

In the lacustrine habitations of Switzerland, the remains of the greater common sheep, with horns curved backwards and then turned forwards, appear in the epoch of transition; in the *marl earth* we have been unable to find stems of such horns.

Wild Mammalia.

We cannot hope to excavate from our *marl-beds* many remains of savage animals. The use of arms in metal and the progress of pastoral life had already widened the circle of savagery. We may, certainly, in time recover the remains of other species besides those of the smallest, the remains of which are disinterred at the present time; but we shall never reach the number of species represented in the pile habitations of Switzerland, some of which were in existence in the age of stone. If, owing to this reason—of the later epoch of our *marl-beds* in comparison with the pile habitations—it appears improbable that we can ever find in those earths the remains of the buffalo (*Bos urus, L. primigenius, Boj.*), of the bison (*Bos bison, L. urus, Boj.*), and other species, living at the same time in more remote epochs with the inhabitants of the Swiss pile habitations. Reasons of topographical distribution would persuade us that we should find with difficulty the remains of the wild goat, chamois, elk, and ermine, which are, however, dug up from the remains of the ancient habitations of the Swiss lakes, though they are in those localities rather rare. We might, on the contrary, presage, with some chance of hitting the mark, that, should we extend our researches, we should bring to light evidence of the existence of the following species in the age of the formation of the *marl-beds*, the cat, the otter, the martin, the pole-cat, the weasel, the civet-cat [?] (*puzzola*), the wolf, the fox, the badger, the hedge-hog, the squirrel, the hare, the beaver, and the fallow deer.

Every one knows that certain species, as for example the common rat and the rat of the shambles (*Mus decumanus*), have come into and been acclimatised in our country in recent times; it is unnecessary, therefore, to say that we shall not meet with their remains in the *marl earth*; in the contrary case we should have the proof that they inhabited these places at the time of the formation of the earth in question, afterwards disappeared from them, were either destroyed or retired into countries more propitious to them, and have since made their appearance again.

The four species at present recognised, from the remains of the *marl earth*, are the stag, roebuck, boar, and bear.

---

* Gibbon, speaking of the Lombards, says, "That they altered and improved the race of animals, especially horses and oxen, introducing wild horses (*Caballi sylvatici*) and buffaloes. Tunc primum A.D. 596. *Bubali* in Italiam delati Italiæ populis miracula fuere. (Paul Warnefrid, l. iv, c. ii.) Gibbon seems to think if buffaloes, in fact, they have come from Africa or India; but adds, "Yet I must not conceal the suspicion that Paul, by a vulgar error, may have applied the name *bubalus* to the auerochs, or wild bull of ancient Germany" (vol. viii, p. 151, n. 44). Probably this animal, which Gibbon imagined was the äurochs, was the *Bos frontosus* afterwards mentioned. Was this Cæsar's *Urus?*—EDITOR.

THE STAG (*Cervus Elaphus*).

If from the length and strength of the horns and their branches we were bound to argue the size of the stags who bore them, it would be clear that these were of a stature above the common ones actually living in Europe. Here are the measurements of the larger horns, expressed in millimeters:—

Total length	600
Diameter of root	78-90
Diameter of the spreading of the crown (*della palmatura della corona*)	175
Length of principal branch	265-295
Length of a shaft (*fuso*)	320

At the sight of them, Signor Mortillet allowed himself to be induced to accept them as of the *Cervus megaceros*, and to declare them to be so to the Italian Society of Natural Science in Milan, at their meeting of the 28th of July, in the present year. We, following Rütimeyer, are unable to perceive in the remains of the stag of the *marl-beds* anything except a large variety of the common stag, *Cervus Elaphus*, which, with the stature of a horse, is the mean between the *Cervus megaceros* and the common living one, and has horns with stem and branches, with a crown more or less palmated, exactly like the horns of the stag of the palisades of Switzerland, and the drift (*diluvio*) of Russia: and certainly the remains of this same stag, *C. Elaphus fossilis*, Goldf., or the *priscus*, or *primigenius*, or of the cited *megaceros*, are also found in our drift, together with those of the buffalo and rhinoceros, *Rhynoceros leptorhinus*; but with the exception of the remains of the first stag, they are not found in the marl earth. We have remarked before that the bones of the stag were chosen for the preparation of utensils and arms, and on that account it does not excite surprise that so few should, up to the present time, have been excavated from the *marl-beds*;* and this fact

---

* Measurement in millimeters of some parts of pieces of bone of stag:—

	Living.	Of the marl-beds.	Lake Habitations.
Metacarpus: greatest diameter across superficies of superior articulation	33	39	42
Metatarsus, id. id.	30	38	38
Tibia, id. lower	37	44	50

is also less surprising because, for the reasons given above, the quantity of that game must have been already much diminished at the time of the formation of the said earths; we have been informed, on the other hand, that from the remains of the Swiss lake-dwellings of the first age, the bones and horns of this species compose the greatest quantity. Rütimeyer observed, that while the skulls of stags in the palisades were like the others, deprived of the facial bone, the cranium had not been opened in the same way to empty out the brain, but so as to present the bony cavity of the brain split open in the middle; while in the skulls of pigs and oxen they appear opened by the removal of the temporal bones. This artifice seems to have been used for the reason that the arch of the skull, owing to its considerable thickness and strength, opposed much resistance to fracture; on the other hand, the only piece of frontal bone found as yet in the *marl-beds*, appears to come from a skull split in the usual manner. For the history of the ruminant in question, the fact is not without interest, that of the large variety of the *marl-beds* and turbaries, remains are discovered in the deposits of the Roman epoch.

The Roebuck (*Cervus capreolus*).

Rütimeyer, of whom we have spoken above, has established that, in the lacustrine habitations of Switzerland of the age of stone, the remains of this beautiful stag show themselves; on the other hand, they are wanting in the age of bronze, and they appear again, only later, in the deposits and tumuli which date from the sixth century after Christ. Wishing to make use of this fact and apply it to the marl-beds, we should be induced to declare them of the later Roman epoch, since we shall see that they are certainly not of the stone age. However, such an induction would be erroneous, since reasons and facts which we shall adduce in the sequel, lead one to hold these earths not to belong to the modern time alluded to. The bones of the roe-deer, up to the present time, excavated from the *marl earth*, are few,—pieces of humerus and femur, radii, tibiæ; nor are the horns many, though of every age,—stems with two and three branches; the largest horn of an adult measures 250 millimeters in length to 45 of diameter at the root. From these proportions, and those of the few bones,* we may infer that the roe-deer of the *marl-beds* was different, perhaps, both from the living species and from that of the drift (*diluvium*) in being smaller. In the form of the horns we find no distinction.

THE BOAR (*Sus scrofa ferus*).

Of this dreaded wild beast we have only one piece to depend upon, with which we had been favoured by the engineer Berté Eugenio, with the remark that it had been taken from the *marl earth* of Castione. The state of preservation in which it is persuades us that it is of recent date; and without wishing to deny its derivation, we are compelled to hold that it was excavated from the upper strata of the said quarry of *marl earth*, which are alluvial, or else that it was only by accident found within the inferior strata of ashes, carbon, and azotised substances; in its greatest curvature it is 220 millimeters, having a greatest diameter of 25 millimeters. From the fragments of remains dug out of the marl-bed of Vico-fertile, there are also some of the bones and tusks of pigs, and we may suppose that among them there are some of the wild boar. The boar of the Swiss lake habitations, common in the age of stone and that of bronze, as Rütimeyer tells us, exceeded in size the greater living ones.

THE BEAR (*Ursus arctos*).

One canine tooth, which came from the marl-bed of Campeggine, is the only trace which guides us in the marl earth in regard to this wild beast. From this circumstance, and from

---

* Some measurements of bones of roe-deer of the *marl-beds*:—

Humerus: breadth of trochlea	29	mill.
Radius, total breadth	145	,,
,, breadth of superior superficies of articulation	29	,,
,, idem, inferior	26	,,
,, least diameter of diaphysis	17	,,
Tibia, total length	140	,,

its being well preserved, we conclude that the animal to which it belonged fell only by accident into the hands of the people

Fig. 26.

of our marsh habitations, who only kept its remains as trophies, or charms, or relics, and not that the bear inhabited our plains at that time; and we assert this in analogy with what Rütimeyer has observed by the study of the remains of the bear of the lacustrine habitations of Switzerland. The bear's tooth remarked on above is 85 millimeters long, and 26 broad, and we consider it to belong to the common bear, rather than to the diluvial bear of the caverns, *Ursus spelæus*.

Birds.

A right tibia of a gallinaceous bird (*pollo*), 104 millimeters in length, was disinterred from the *marl-bed* of Castelazzo di Fontanellato; and a piece of a right *humerus* of the same bird was found in earth of the same kind in making an excavation within the walls of the city of Parma; also among the remains of the pile habitations of Morges in Switzerland, supposed to be of the bronze age, was discovered a hen's bone; but Rütimeyer recognises in it the remains of a recent epoch, and denies that it can be of that age. As regards the relic from the *marl-bed* of Castelazzo, we may asseverate that it was taken out of ashes and carbon, which in part adhered to it closely, and so we may hold it certain that it was extracted from the stratum of the *marl earth*. With regard to the other bone of the common fowl, it was found mixed with bones of the small ox of the *marl-beds*, and of other mammalia, with shards of Celtic earthenware and charcoal. The state of fossilisation of these remains of birds, above all of the tibia, does not permit us to regard them as of recent date.

In time we may hope to find, also, in the *marl earth*, the traces of marsh birds and reptiles, inasmuch as we have found there the shells of freshwater molluscs, a list of which we shall presently give; the first, also, have not been wanting in giving their due contingent. Rütimeyer enumerates eighteen species of birds belonging to the orders *Rapaces*, *Passeres*, *Paperine*, *Gallinæ*, *Pinnatæ*, and *Natatores*, besides three species of reptiles, and nine of fishes, the remains of which have been dug up from the Celtic pile habitations of Switzerland.

Molluscs.

Underneath is the list of the shells collected in the *marl-beds*, in every respect like those of molluscs now living in this country. Let us hasten to add, that some of them do not actually live in the plain, but only on the hills or at their feet, whence we may hold that their shells have been transported down by the torrents, and perhaps also were deposited by them; while the aquatic shells, at least of the *acephali*, were deposited by the Po, which bathed and inundated periodically the habitations of those ancient and rude peoples.

Aquatic Molluscs.

Acephali:—
    Unio pictorum; var. Requienii. Mich.
    Alasmodonta compressa. Menke.
    Anodon. (In pieces.)

Gasteropoda:—
    Paludina vivipara. L.
    „   achatina. Lam.
    Limnœus stagnalis; L., var. minor.

Land Molluscs.

Of the Plain:—
    Zonites Draparnaudi. Beck.
    Helix Carthusianella. Drap.
    „   hispidi. L. var.
    „   nemoralis. L.

Of the Foot of the Hills:—
    Helix nemoralis; L. var. etrusca. Ziegl.
    „   lucorum. Müll.
    Cyclostoma elegans. Müll.

Of the Hills:—
Zonites olivetorum. Müll. Var. Leopoldianus. Charp.
    Helix obvoluta. Müll.

Vegetable Remains.

Except a few roots, some indeterminable seeds of some sort of *Hypnum,* and an acorn (*ghianda*), about which remark is made in the special account of the pile habitations of Castione, we have not been able, up to the present time, to find in the *marl-beds* vegetable remains of any kind. However, we have been assured by various persons worthy of credit, that within this earth is found, pretty frequently, *corn,** *vetch,* and *bean,* carbonised and heaped together; and you yourself mention this in your Memoir, from which we suspect that these may be the remains of magazines of these cereals. One person asserts, that there have been dug up in the *marl-beds pear* and *fruit stones* completely carbonised.

To complete the very short and imperfect remarks upon the vegetable species represented among the remains of the *marl earth*, as far as present results go, we shall allude to the *chestnut* and *elm,* of which wood the piles, which have been alluded to, are roughly hewn.

---

* *Triticum turgidum, L.*

Concluding Remarks upon the Organic Remains.

Now that we have completed the review of the few organic remains of the *marl-beds*, it is convenient to examine what sort of changes, in the natural productions in discussion, have occurred on the theatre of these fertilising deposits from the disappearance of the populations, to which we owe them, to the present day. Owing to want of material, we are obliged to confine ourselves to the *Fauna* alone.

We have not been able to perceive any alteration in form in the shells of the molluscs; there is no difference whatever between the shells found in the *marl earth* and those which the molluscs of the same species secrete at the present day. From having found the shells of molluscs which at the present day inhabit the hills, while nearly all the *marl-beds* are situated in the plain, it does not follow that we must deduce that these soft animals then lived in the level ground, but, as we have already maintained, that the spoils of the molluscs, as long back as that time, were brought down by the waters of the Apennine.

Of *birds* we have only collected the bones of a single species; from their marked features, we might argue that the fowls to which they belonged enjoyed an independence which we are not accustomed to accord to them. It now only remains for us to follow on the traces of the changes which have happened in the fauna of the mammalia, whether by modification of their forms, the change of the domestic races, or by the alteration or disappearance of them or their stocks, or of the wild species, and by the examination of these animals we may expect some interesting results.

The pig, the ox, and the sheep of the *turbaries* and *marl-beds*, among the domestic races; the stag, the roe, and the boar among the wild species, have *abandoned* the scene of the marl earth,—the wild animals only in historic times, and even recently. We have already mentioned our opinion, that the bear was only an *accidental* guest of the marsh habitations of our plains; now he is so pushed back and straitened in his native mountains that a similar incursion would be quite impossible.

The dog has undergone some slight *modification* in form, and, thanks to crossing of breeds, is subdivided into an infinity of races. The horse, the pig, and the common bull continue prosperous, almost unaltered in form, but, except the bull, increased in stature; while, on the other hand, the stag and the boar, where they still live in Europe, are diminished in size. The horns of the stag and the tusks of the boar are less than the analogous remains excavated from the *marl earth* and the pile habitations of Switzerland. And this modification in an opposite sense,—in one case progressive, in the other, retrogressive, according as the animal is domestic or wild,—we may observe generally in all the mammalia, save a few exceptions, and is easily explained, by admitting an increment in the conditions favourable to life and prosperity for the domestic animals, and a decrement of the same circumstances for the wild ones,—a victory of pastoral and agricultural life; and so of human civilisation.

The wild stocks of the pig of the *marl-beds* and of the two bovine races, descended from the diluvial epoch, are actually *extinct*, as far as appears from diligent zoological and geological research.* We find continually the traces of the pig, the ox, and the sheep of the *marl-beds*, but domesticated, near the theatre of the pile habitations of Switzerland, in the canton of the Grisons, as well as elsewhere in Europe, chiefly in the north.

Of the *domesticated* species, the small ox was the first brought under man in Europe; next the dog, the goat, and goat-horned sheep,—all in the age of stone: next followed, in the same age, the common bull, the small pig, and the horse; the last to enter into the family of domestic mammalia

---

* The above wild pig disappeared, as we learn from Rütimeyer, from the theatre of the lake habitations of Switzerland, with the introduction of the arms in bronze, certainly of more use for hunting than those rude ones in stone; and as further, bronze, and generally metal, lent itself better than stone to the fabrication of implements and instruments, domestic, agricultural, and industrial; so their introduction giving a strong impulse to progress, contributed also indirectly to the limitation and destruction of savage animals in general.

of Europe was the pig (*majale*), as appears probable, in the age of bronze.*

We shall end this zoological epilogue by adducing other facts observed on the subject by Rütimeyer,' besides those already mentioned, and at the same time setting out the deductions which he draws, persuaded that we shall in this way please you, and be advantageous to many persons; because it will make them acquainted with a recent work of value, which, owing to its being written in the German language, cannot be consulted by many, because that tongue for the most part, and very undeservedly, is little studied by us; especially from motives little plausible, but still pardonable, the cessation of which, we hope, is imminent, with the cessation, in our country, of an administration which has made use of, and even makes use of, the torch of national hatred,—a hatred the most irrational, to prolong its hybrid existence.

Rütimeyer divides the time, since the foundation of the first lacustrine habitations in Switzerland, into *Three Ages*.

First Age,
OF THE PRIMITIVE DOMESTIC RACES.

In this period, the wild animals used for food (*alimentarii*) far surpass in numbers the domestic; the chase preponderates over pastoral life. This epoch coincides with the age of stone of the antiquarians.

Second Age,
OF THE MULTIPLICATION OF THE DOMESTIC RACES.

This is not separated from the preceding by well-marked

---

* The common pig (*majale*) does not appear to have been domesticated in the place of the Swiss lake dwellings, but to have been brought there tame. A great race of dog, nearly related to the butcher's dog (*cane da beccajo*), and a variety of the pig of the turbaries, still less than the type, both perhaps, imported into Switzerland, mark there the beginning of the age of bronze. The pig disappeared there towards the sixth century, perhaps for want of recruiting from without; we have neither found in the *marl-beds* the remains of one nor the other, but only of the bones of the pig of the turbaries, or *marl-beds*, or pile habitations,—whichever you like to call it,—which in measure stands between the type and variety in discussion, and molar teeth, which in form are suitable to the same variety.

Cane da Beccajo, I believe, answers to our bulldog, and is the *Canis familiaris molossus* of Linnæus.—EDITOR.

limits, and it is less still from that which follows it; the domestic races of the first epoch pass into this, and their brains and marrow are equally made use of. Two races, however, disappear during this epoch, and in their place four others present themselves on the scene; while, on the other hand, two wild species die out, and on this account the number of domestic animals used for food begins to preponderate over that of the wild ones. The introduction of metal marks this epoch; however, this age does not correspond with the age of bronze, as it gradually confounds itself with the existing age.

THIRD AGE,
OF THE CULTIVATION, AND THE PERFECTING OF THE DOMESTIC RACES (KULTUR RACEN).

Or rather, the existing age, characterised in Switzerland by the appearance of a new race of oxen, *Bos frontosus* (*fleckvich*). Wild animals become the food of luxury; on the other hand, cattle increase, the races multiply and become adapted to various climates, uses, and wants,—thanks to breeding and crossing of breeds.

This age begins, probably, with history; to the two preceding ones we must assign a somewhat long *duration*, not, however, after the manner of the geologists.

The discovery of pile habitations and objects in stone and bronze, of pre-historic character in Italy, the theatre of the Roman life, promises, according to Rütimeyer, a not distant elucidation in the matter, and we are rejoiced to be able, in some degree, to assist such elucidations.

The domestic animals of the more ancient periods, such as those true and pure primitive races, were in part in a state of liberty, already the inhabitants of the same regions of Helvetia, but they had come from other regions, probably from the south; so that the domestic *fauna* of the first age, although of the land, was not indigenous. The domestic races of the second age would seem to have been imported, and to have died out for want of being recruited. The ox which in Switzerland marks the second age, appears to have arrived from the north. These migrations of herds must place us, some day, upon the traces of their leader and shepherd, man; and to be able to follow the path of the latter, it is necessary to know the fauna, present and past, of the different countries.

By the comparison of the fossils of the turfy lignite (*lignite torbosa—shieferkohle*) of Dürnten, and the diluvial and glacial deposits of the valley of the Rhine which rest upon it, with the remains of the turf which conceals the lacustrine habitations of Robenhausen, Rütimeyer deduces that at the time of the formation of the lignite in that country, lived at the same time the stag and urus, with the elephant and rhinoceros; afterwards, in the period of the depositing of the drift (the upper part of it), with the preceding animals were associated the mammoth, the wild boar, the rein-deer, the roe, the wolf, the fox, the badger, the beaver, and the hare. This fauna was afterwards driven back, and in part expelled from the valleys of Switzerland and from the country by the *slow* increase of the *glaciers*, which, to the rein-deer and mammoth alone allowed a stinted existence. Afterwards, when by slow degrees these enormous deposits of snow and ice retired, the gigantic ox (*urus*) reappeared in Switzerland, and by *little* and *little* ascended the valleys, accompanied by the bison and the elk, but no longer by the gigantic pachyderms, which were their fellows in the preceding age. In this epoch, which is that of *the reappearance of vegetation,* falls the arrival of man in Helvetia. From that time forth there was *not* room for *great changes* in the fauna of that country. Though the disappearance of the rein-deer, and that of the bison and the urus, happened in historic times, as did that of the elk, the beaver, and the stag, we should not, perhaps, attribute this entirely to man.

It may be that the same causes contributed to it which limit within climates—certainly, at present, more cold—the chamois and the wild goat, not to mention plants entirely independent of man, which organisms prospered on the theatre, now certainly more mild, of the lacustrine habitations; on the other hand, moreover, only agencies connected with climate could have united, within the same space, animals which at the present day are separated by half a meridian, the elephant and the rein-deer. Already in Cæsar's time, who describes the urus and the elk of the Hercynian forest, these ruminants were separated from the elephant by a distance of 30° or 40° of latitude. We know, too, that in the present geological epoch there are signs that animals attempt, as is the case at present in the tropics, to penetrate from the south towards the north. Among the remains of the lake habitations of Switzerland we find the horns of the fallow deer; and we know that in the historic age the lion has appeared in Greece.

With man trod the Helvetic soil, now abandoned by the ice, various species of domestic animals, some perhaps subdued on the spot, others, and *with greater probability*, conducted with them in a state of domestication. The lacustrine people of the palisades do not seem, then, to have been the first *inhabitants* (*dei primi abitatori*) of Europe. These their water villages are probably the most ancient storehouse of the life of man in Switzerland, because those anterior to them, if any such existed, were destroyed by the glaciers. But these are not the first traces of man in Europe. In fact, elsewhere, at Aurignac in France, human skeletons have been dug up confused with bones of the mammoth and rhinoceros, and the other precursor animals,—eye-witnesses and successors of the glacial epoch,—bones which were broken like those which have been dug from the pile dwellings and *marl-beds*, and with the same object, and which have been gnawed by the teeth of the tiger and the hyæna of the caverns; elsewhere, in France, have been found remains of the elephant and rhinoceros, with monuments of the most ancient art of man.

Man, therefore, lived in Europe before the glacial epoch, during the depositing of the drift. With these discoveries the barrier is overpassed, which existed till now, between time past and present,—between geology and history.

Conjectures on the Origin, the Age, and the People of the Marl-beds.

Upon an acquaintance with the organic remains and the pre-Roman monuments buried in our *marl-beds*, the question at once presents itself, by what, and what kind of agents they were accumulated? for what object, and in what epoch? We do not pretend so far as to answer these questions categorically; we shall be contented with passing in review the various opinions which we know to have been ventured upon this subject, with criticising them, and finally, advancing an hypothesis in argument, as far as we may be able, after one half-year only of research and study, sometimes at the mercy of circumstances, sometimes of the will of other people.

Some of these conjectures have been combated by you in the memoir which has been cited, as those of Venturi and Cavedoni. Among the true *marl earths* we only include the first category of the *cemetery earths* of Venturi, the more ancient, those which he holds to be a sort of cemeteries where the Galli Boi buried their warriors, dead in battle; or else the remains of booty burned by them, or of piles and sacrifices of the same age; those earths, in fine, where we find the Celtic potshards, and we exclude the earths which contain Roman burial places *only*, and fabrics destroyed by some fire, or rather ruined.

Cavedoni, as you say, remarking upon the apparent stratification (of part, at least) of the *marl-beds*, shuts out the probability that they could have been cemeteries of bodies buried in haste (*tumultuarie*), or of prey alone, or even once or twice *burnt* on the spot.* Signor Mariotti,† speaking of the *marl earth* of Marano, demonstrates—as too strange, not to say specious—the supposition that a deposit of human corpses extended to the east from Marano to Monticelli, and extended many chilometers (about six) along the towns of Malandriano, Coloreto, and S. Prospero, and perhaps also along the lower towns, at least to the mouth of the Gambalone river in the Enza;

---

* According to Ricci, *Corografia dei Territorii di Modena, Reggio ecc.*, Modena, 1788, the booty was not burnt, but, according to the testimony of Cæsar, heaped up only; these accumulations, sacred to the divinities, remained untouched, owing to religious respect for them; we think that he meant that this booty was heaped up, perhaps sometimes burnt as well.

† *Notizie intorno la canaletto di Mamiano e Monticelli raccolte da Gaetano Mariotti. Inedite.*

and we may add, that in the greater part of the *marl-beds* are found scoriæ, sometimes abundant, which certainly were not buried with the warriors, nor do they come from the burning of their arms or harness, but were, as we show elsewhere, the remains of a forge little disturbed, where are gathered also the moulds for running the arms and instruments; we should also reflect that the arms dug up are few, while the custom of these people is known of burning or burying the warriors with their arms. The extreme difficulty in meeting with human bones in the strata of the *marl earth*, also alluded to by Mariotti, would be a further ground for doubting a like origin with these earths, in which it is agreed human corpses have been buried.*

Against the hypothesis of the cemeteries and the burning of booty stands further the fact, that we find there a great quantity of bones of animals broken to pieces, but not burnt; and that we collect in incredible abundance the shards of vases of all kinds, and not only of those which we can consider as sacred. Finally; to burn prey and bury corpses, extensive palisades are not planted, grain is not heaped up, nor millstones arranged for grinding corn. And these same facts speak in great part also against the other opinion of Venturi, embraced by Cavedoni, that these *marl earths* are the remains of funeral pyres and sacrifices of the Boi. A similar opinion is, as you know, in like manner sustained by Ghiozzi in his memoir on Giulia Fidenza,† where, speaking of the marl-bed of Bargone di Salso, he says, "it is proved that the Etruscans met there, the Galli Anani and Boi, and afterwards the Romans, to make their sacrifices." But in this *marl-bed*, also, we found, as well as Gramizzi, pieces of earthenware by thousands, of every kind, scoriæ, and millstones. We come now to the hypothesis, several times mentioned in your memoir, which is this, "that *the greater part* of the *marl-beds* are

---

* That is, with Venturi's other cemetery earths mentioned above, from which the first category is considered distinct and of different origin.—EDITOR.

† *Memorie storiche sulla fondazione della città di Giulia Fidenza*, edizione seconda. Borgo S. Donnino, 1840.

remains of Roman sepulchres and remains of cemeteries, of pyres, and perhaps, also, of convivial meeting places of the bronze age, *rearranged by the action of waters."*

We have already excluded from the *marl earths* those earths which contain the Roman burying places; however, we shall again, in the sequel, return to this discussion, as we ought to give our reasons for excluding them. To the conjecture of pyres and cemeteries, we have already given the objections which have appeared to rise in opposition. With regard to the idea of cemeteries, granted that the bodies of the people of the *marl-beds* were buried, rather than burnt, which we do not concede, for reasons which we shall soon give. There appears to be a proof in support of our opposition in the fact accepted by you, that at Cumarola was discovered the cemetery of forty warriors, with arms of stone and bronze at their side, a cemetery which you hold to be of the same epoch as, though not of the very formation (*giacitura,* laying down) of, the *marl-beds,*—that is, of the age of bronze. There is no mention made by the discoverer either of ashes, or charcoal, or shards, or scoriæ, or millstones, nor of bones of animals, etc., objects which are all found in the *marl-beds;* so that cemeteries which are granted to be such (*in posto*) do not give rise, as Venturi supposes, to these azotised deposits. No more can cemeteries turned *topsy-turvy* (*sconvolti*) by water give place to their deposition, since we do not meet there with the components of the *marl earth.* The people to whom these deposits are due, do not seem to have had the custom of burying the dead, but certainly of burning them; since in the marl earth the bones of men are of very rare occurrence, which belong to the age of the formation of those deposits. On the contrary, the people to which the warriors of Cumarola belonged, used to give their sepulture outside the enclosure of the habitations; just as Rütimeyer holds the inhabitants of the lake habitations of Helvetia used to do, and, according to Bertrand,* certain tribes of Gaul. We therefore suppose that these and the said warriors belonged to a single nation, the

---

* *Revue Archéologique,* Jan. 1861, p. 9.

members of which lived upon both sides of the Alps,—that is, in the beginning of the period of transition, and in our opinion, anteriorly to the people of the *marl-beds*, which had, in part, different habits from them. You hold, as has been said, that these earths are the remains of banquets, and we agree with you. You suppose that these earths have been altered in position by water, and we fully admit it; but we do not venture to say this for the *greater* part, for we have not visited more than about a third of those which have come under our cognisance.

The visible stratification, or, we should rather say, supraposition of beds—sometimes curved and sometimes horizontal—of *ashes* with carbon, alternating with deposits of earths, as are observed especially at Casaroldo di Samboseto, Conventino di Castione Collechio, Fodico di Poviglio, and Vico-fertile, indicate that these deposits are in position, or rather in the place where, in a long succession of centuries, they were formed by the work of man, followed by physical agencies. At Casaroldo, Castione, and Poviglio, the placid overflowings of the Po have deposited in part the strata of earth over the various beds of ashes and charcoal. We are led to believe this by the shells of the acephalous molluscs in contact with these substances, as these molluscs always live in the waters of the Po. In the places named, Collecchio, Vicofertile, and elsewhere, the inundations you mention of the torrents from the Apennines, *perhaps*, added the snail molluscs of the hills and the little tertiary fossil-shells. It is possible, then, that these, transported by the torrents into the Po, may have been brought back over these deposits by its waves when that river, as is probable in such remote times, covered our whole plains, and it is here that almost all the *marl-beds* lie which have been observed; we do not know one situated on a hill, only at the foot of the hills one has been found.* As to the

---

* The number of *marl-beds* visited by us amounts to twenty, there remain still to be seen about forty, of which we have knowledge; those more to the west, known to us, are those of Besenzone, Castione, and Salso, in the circuit of Borgo, S. Donnino; and those most to the east are those of Castelvetro and Nonantola, in the Modenese. We understand there are also some

G

marl earth of Castione, the palisades, extending under its position, prevent one from having any doubt that the earth is *in situ*.

From these heaps then in position, the rain-waters—those of rivulets and brooks, the currents of rivers, the waves of the Po more or less gently, sometimes, however, with impetus and precipitously—have carried down, to a greater or less distance, parts of the earths with their contents, or have only eaten into them and turned them over. In both these last cases, however, we cannot find the beds of ashes and charcoal, since the waves will have dispersed them; the shards, moreover, are without corners and pebble-shaped, which is also asserted concerning the marl-bed of Marano, considered to be transported by the above-named Sig. Mariotti.* Here is his criterion for distinguishing the *marl earth* in position, or *virgin*, from the washings, whether they lie bordering upon the same earth or ridges (*lembi*), or whether they be constituted apart from them in separate deposits or alluvial beds (*alluvioni*); such as are the marl-beds of Basilicanuova, Marano, Beneceto, Gajone, Madregolo.

The first, the virgin earths, as they have been called, include *beds* of ashes and charcoal: the marl earth is usually more clear, more bright, calcareous, azotised, and fertilising,—it is *marl earth* of the *first quality*. In the second, whether it be in ridges or alluvial beds, *ashes* and *charcoal* cannot be discerned, they are *spread abroad* and dispersed, and are therefore wanting; for example, at Gajone. The colour of the *marl earth* tends to brown; it is argillaceous and more heavy, less fertilising; it is of *inferior quality;* it is also rotten, and does not repay the expenses of excavation and transport.

As to the first *origin* of the *marl earths*, it is clear that the

---

in the Ferrarese. The *marl-beds* nearest the Po, and therefore most liable to inundation, are those of Poviglio, Brescello, Casaroldo, and Besenzone; near the hills, those of Castelvetro, Pontenuovo, Torrecchiara, Collecchio, and Salso.

* In the alluvial *marl-bed* of Marano, according to Mariotti, bones are not found. It would appear, according to Mariotti, that this *marl-bed* has been an island of the Parma.

banquets, as you assert, are a considerable part; but there seems to us to appear in the scoriæ, the millstones, the heaps of grain, the palisades, the potsherds, already cited, together with the arms and utensils of all sorts which are found in these earths, something more than a mere meeting place to banquet. It seems to us, if we do not err, that there is something of settlement and duration. Man did not meet there only to arrange and devour the feast; but to employ himself besides in domestic avocations, in preparing implements and arms, to sew garments and make nets,—in a word, to inhabit them; besides to exercise the practices of their religious worship, and perhaps, also, to burn their dead, and all these after the fashion of barbarians, such as the people of the *marl-beds* must have been. These people, according to the place and time, were fishermen, hunters, shepherds, and even agriculturalists. For fishermen was required a safe residence, and defence against the periodical overflowings of the waters; against the assaults of wild beasts and enemies, was wanting a pile dwelling. The hunter and the shepherd have no other want than of a high and dry station, where to plant their movable tent, to which the nomad herdsman returned, when with his flock he appeared again upon the pastures which surrounded it, and when he abandoned anew, in turn, other stations and other pastures, either known already or new ones. The agriculturist was settled upon the land he cultivated, and had a more solid habitation of clay or dry wall, and strongly made. Among the remains of the marl beds we find not only the pile dwelling and dry wall, which attest the different conditions of the people in discussion, but we find also the hooks which indicate the fisherman, the horns of stags, the tusks of the boars, the teeth of the bears which the hunter pursued, and the bones of domestic animals, which distinguish the shepherd; finally, the scythes, and the magazines of cereals, which show that they were also agriculturalists.

While our plain was increasing by little and little from the level by fluviatile deposits, the marl earths rose up also by the hand of man, or rather, for the most part both by that and by the action of the waters, only that the accumulation of the deposits of man went on less slowly; whence arose the forms of mounds (*rialzi mammellonari*) of marl earth in position. And on these hillocks the man of the present builds his habitations, his villas, or else convents and castles, as if he wished to approve in this way the convenience of the selection of place made by his rude ancestors, who from their mounds defied the elements and the wild beasts, spied out the enemy, and made ready either for attack or defence; and if man at the present time holds his domicile there, what wonder that our Roman fathers should have taken up their abode there also, their first inhabitants being driven out or subjugated. This is why we find so many Roman memorials in the *marl earth* in position, but either only in the upper strata of the mould or alluvial deposit, or, as is sometimes the case, in the *marl earth* itself, but placed in it posteriorly; as for instance, the buried corpse. The same reason accounts for our finding on the surface so many objects of the middle ages. In the alluvial *marl-beds* we find remains of every epoch mixed together, as you have observed in your memoir;* and since it appears, up to the present time, that the Roman and later remains do not form an essential part, but only an accidental one, of the *marl-beds*, we pass them over in silence; and although the corpses and other organic remains of the Roman or a later epoch may generate, where they are buried, bands of earth a *little* azotised and fertilising, still as they do not enter into the conception which we have formed of the *marl-beds*, we exclude them.

The *origin* and *position*, or rather the mode in which they were formed and deposited being established, it is fitting to inquire *whence* came the *nation* which has left these relics, and at what age it peopled this country? We shall first try to make it out taking for our sole guide the remains of their

---

* Sig. Mariotti, remarking on the Roman objects found in the marl-bed at Marano, such as pans, spoons, hafts, and hilts of poignards, in bronze, coins, tiles, says, that he collected them chiefly in the upper strata. Holding this *marl-bed* to be alluvial, we shall, by such a fact, be led to the supposition that it was transported and deposited before the Roman colonisation, and afterwards increased during that age.

domestic animals and the monuments of their industry; in the sequence we shall interrogate history.

We have seen how they possessed, in common with the people of the pile dwellings of Switzerland, certain races of dogs, of sheep, of oxen, and pigs; it would seem from this that they might have come from thence; but in fact this should prove just as much that the people of Helvetia had already trod the soil of Italy, as that they, transcending the Alps, had entered from thence, although the fact that the goat-horned sheep, the ox, and the pig of the turbaries, no longer exist in the country of the marl-beds, while some traces of them now remain in Switzerland, would support the first assertion. However, we have in this, at least, evidence to lead us to hold that the inhabitants of the palisades of Helvetia and those of ours were of a single stock (*stirpe*),* as you and Keller have already admitted.

You have yourself already noted that in the *marl-beds* we meet with pieces of millstone of garnet-bearing talco-schist; and we, too, collect them, together with pieces of millstone of mica-schist, also garnet-bearing, both stones which, to our knowledge, are not found on our Apennine, and still less likely to come from our torrents in the plain.† You have reported already in that mica-schist a perfect resemblance with the same rock in position in the most ancient coppermine of S. Marcel, in the valley of Aosta, without exactly limiting to this locality the fabrication of these millstones of

---

* Not, however, of the same nation. We have already noticed some differences both in the mode of emptying the brain and the teeth pulp of the mammals, and in the mode of paying the last honours to the dead.

† Even if these rocks existed in the mountain, the people of the marl-beds would, nevertheless, have had them with difficulty, since, as we have said, they remained in the plain and low hills. In the mountains there is no trace of them, and it would seem improbable, not to say impossible, that the waters should have carried down *all* monuments of them; to which conclusion we are naturally brought if, with Sig. Mariotti, we were to admit that our mountains were inhabited by the people of the *marl-beds* (before the drying up of the plain in the valley of the Parma). They knew, certainly, how to find steatite and limestone in the beds of the torrents for spindle-whirls, and to procure from the low hills (*colle*) other stones for the construction of millstones.

mica-schist and the production of copper, objects which are wanting in the region of the *marl-beds*. It seems that we may admit, in consequence, that they come either from there or from the other valleys of the Western Alps, where these minerals are found in position. As to the copper, certainly the people of the marl-beds may have obtained it in a commercial way, but as to the millstones in schist, this would not appear likely; for as there were in position other stones which could be substituted for it with advantage, this would not have been any good to them; and, in fact, we see that the pieces of millstone in schist, excavated from the *marl earth*, are few in comparison with those in macigno, and other stones which take a polish, in breccia and serpentine,—rocks which are met with at the foot of the Apennines. From all this we should infer that the millstones in schist, which are met with in the *marl-beds*, were brought from the *Alps*, either in the west or northwest, and since then, according to the researches of Keller and yourself, as well in the turbaries of the lakes which drain both sides of the Alps, as well as elsewhere, we discover traces of the same *race*, we conclude that the people of the *marl-beds* came from them.

In the *marl earth* are found very few arms and very few utensils, and other objects in *stone;* on the other hand, many more are collected in *metal*.* In the different pile dwellings of Switzerland, on the contrary, we find only arms and implements in stone (mixed, be it understood, with those in bone and wood, common to either time). The remains of the turbaries of Italy belong, as you affirm, to the age of *bronze;* and, as we have already admitted, that the people of the *marl-beds* of Emilia, and that of the pile habitations and turbaries, already mentioned, were of the same stock (*stirpe*), so we recognise in the succession of those epochs the indication of the life which they led in their migrations. Beyond the Alps they dwelt in the age of stone, here only in that of metal; so

---

* The vases with handles preponderate over those without; the first, according to Troyon, *Rev. Archéol.*, 1860, p. 37, are not found among the remains of the age of stone.

it was thence they extended themselves to our plains. In this way, the time of their immigration is *approximately* reached, which happened in the age of metal, and therefore not many centuries before the new era. Here it made a prolonged stay, perfecting itself—in contact with the Etruscan, Ligurian, and Roman elements—in industry and agriculture. One glance at the figures, which you have had executed, of the various implements collected in the *marl-beds*, is enough to convince one of this. The person who carved the ruder objects depicted in bone and stone, could not have executed, with so much good taste, the comb and small wheels in bone given above, nor the greater part of the finer objects in bronze; nor, finally, the spindle-whirls in calcareous stone, brought to such a perfection of polish and regularity, that one is led to suspect the use of the lathe.

If next we try to come at the epoch of the irruption of the people of the *marl-beds,* solely by examining the races of their domestic animals, whose remains considerably preponderate over those of the savage mammalia, we shall gather that such advent could not have fallen except in the epoch of bronze, in the middle of the multiplication of the domestic animals, and in the beginning of the new era; not before the bronze age, for before that time the common pig (*majale*) was not brought under subjection, nor was the common domestic cock acclimatised; not long after the beginning of the existing era, because before then disappeared the small races of the pig and the ox, that of the sheep with goat's horns, and the dog of the turbaries. To such conclusions, at least, are we led, judging analogically from the remains of the lake habitations of Switzerland; and as we must admit that this people did not with difficulty invade our country in a spasmodic way, but dwelt there long enough to accumulate the *marl earths,* we gather, in the *chronological extremes* indicated, the limits of their existence as inhabitants of our valleys, from the age of metal to our era. From that time, the people of the *marl-beds* were either destroyed or subjugated by the Romans, or else were fused with them, and, changing their customs, changed also the construction of their houses, either abandoning the pile dwellings or converting them into strong houses.

Before seeking enlightenment from history regarding this people, we ought without its help to do away a doubt regarding the epoch of their appearance,—a doubt which arises in reading your memoir. You mention, that near Imola were collected, from underground, arms in bone and stone, in part made on the spot, spindle-whirls and millstones, of the beforementioned garnet-bearing mica-schist, and this without any mixture of objects in metal; whence it follows that these arms and implements belong to the most ancient age of stone. The mica-schist would lead one to hold that the manufactures of these objects had descended into the valley of the Po in the same direction whence came the people of the *marl-beds*, and the spindle-whirls and the millstones would lead us to suppose that they both belonged to the same race. And since those remains at Imola are of the age of stone, we are led, hence, to infer that the originals of the people in discourse descended from the Alps, exactly at that epoch; and so that the inhabitants of our pile dwellings came here in the same epoch, while we make their appearance only reach back to the age of bronze. We cannot deny that the manufacturers of the rude arms in flint, of Imola, came from the Western Alps; but we are not inclined further to admit that they were one people with the inhabitants of the *marl-beds*; for the former appear to have had different customs from the latter, and perhaps the same as those of the warriors of Cumarola,—a reason why they did not leave *marl-beds* as evidence of their stay in Italy. But suppose we admit that both belonged to the same stock (*stirpe*); the fact that those of Imola did not know metal, and had in part different customs from the people of the marl-beds, would make us incline to hold that they composed an early colony which crossed the Alps at an epoch anterior to that in which another nation of the same race—a second and greater colony—departed to people the southern plain of Upper Italy.

Now, we may open the book of ancient history, and read in its pages the fortunes of the people of the *marl-beds*.

Titus Livius relates that, in the reign of Tarquinius Priscus, about six centuries before Christ, the *Galli* made a descent upon Italy, invited by the pleasantness of the climate and the deliciousness of the wine. They came in at four several epochs, and in four different bands, the last but one of which was that which comprehended the Galli *Boii*. Micali supposes that these came down by the Mount St. Bernard,* and says, that, rejoiced to breathe an air less raw, they moved on *directly* across the Ticino, and finding the country between the Alps and the Po already occupied by their countrymen, they crossed that river, near its confluence with the Adige, and entered *suddenly* into the region nearest to the Apennine;† a considerable part of that territory towards the Po was always covered with vast marshes,‡ so the Annani made their nest here first; the Boii extended from the Faro to the Idice or to the Sillaro; and in the last place the Fingones occupied all the succeeding tract as far as the river Utente (or Utis), now Montone, in the neighbourhood of the Adriatic. The Senones, coming after them, took possession of the territory between Ravenna and Ancona. From this it may be

---

* According to Contzen, they made their descent by the *Simplon;* whichever way it was, they came from the north-west of the Alps, where exactly we find in position the garnet-bearing schists, of which are composed different millstones of the marl earths.

† Pliny, on the authority of Cornelius Nepos, asserts that the town of Melpo, in Lombardy, was destroyed by the Insubres, Boii, and Senones united, in that day on which Camillus destroyed the city of Veii in Etruria. Lopez draws from this the conclusion that in the year 358 of Rome (396 B.C.) the Galli Boii, precisely those who inhabited this part of the country, were still beyond the Po.

‡ According to Livy, it was peopled by the Etruscans and Umbrians, who were the people invaded by the Gauls. In our opinion, however, the Etruscans and Umbrians either never reached our territory or kept to the mountains, at that time less inhospitable than our plains, since there is no record of their abode here; and we confess that we do not understand how a people could occupy a country, inhabit there, and not leave any relic of themselves. Before the monuments of the Boii, we find no token of human civilisation in our *flat country*, and therefore less any trace of that first epoch of absence of civilisation, by the northern *savants* called the age of stone,—the physical reason we are about to mention; on the other hand, in our Apennine, there are not wanting evidences of this most ancient epoch, as axes, knives, with hatchets (*couteau-hache*), chisels, and hammers in stone.

seen that the country where we find the *marl-earths* is precisely that occupied by the Galli Boii.* Micali goes on to say that the Galli, ignorant of agriculture, lived at that time through the vast forests and marshes (then certainly upon piles, after their custom) in the wretched condition of a hunting and pastoral people. They learnt, but *much later*, of the Italians to cultivate the earth, to divide it, and to possess separately houses and fields, and understood the advantage of uniting the scattered population in durable abodes.

On coming to blows with the Romans, they received the tremendous defeat at Talamone by the valour of the consuls, Cornelius and Æmilius.† After this destruction of forty thousand Galli, the victors crossed over the Apuan Alps (*le Alpi Apuane*‡) into the country of the Boii, or rather into our plain country, and deprived them of about half the territory. In the sequel, the Romans having conceived the idea of completely subduing the Galli Cisalpini, two consular armies entered into the open country of the Boii, and overcame these people; unless, on this occasion, a *part* of them preferred to abandon Italy and go to the north of Noricum and Pannonia, or among the Taurisci,§ and live there near the Danube, where they were afterwards destroyed by the Dacians and the Romans under Augustus.

From the different opinions with regard to the epoch of the decadence, and the total disappearance of the Boii of our country, we adopt those of Livy and Micali, who make them arrive 600 years,‖ or thereabouts, before the new era; and

---

* Pliny, describing the eighth region of Italy, which had for its limits the Po, the Apennine, and Rimini, places there the Boii, and on the authority of Cato makes their tribes number a hundred and twelve.

† This refers to the victory of the consul P. Cornelius, Cn. F. Scipio, 191 B.C.; his colleague was not an Æmilius, but M'. Acilius Glabrio, a plebeian. *Vide* Livy, b. xxxv, 24, 37.—EDITOR.

‡ That part of the Apennines called Apuani Mugelli, from some way north of Spezia, called in Roman times, probably from its extreme loveliness, Portus Veneris, to near Pistoja, then Pistoria.—EDITOR.

§ A people in Noricum, so called.—EDITOR.

‖ Niebuhr considered that Livy's account of the time at which the Galli first came into Italy, about 600 years B.C., in the time of Tarquinius Priscus, was without sufficient warrant. He, however, says, that the legend on

they do not exclude the possibility that their remains discomfited (in 183 B.C.) remained *in part* a long time, in the inhospitable parts of the country, and those shunned by the agricultural and civilised people; such as were at that time the Romans,—their persecutors. There, whether in wild forest or in lagoons difficult of access, concealed from the eye of their enemies, they continued, undoubtedly, in the exercise of the religious worship of the Druids, and the customs of the *Celtic* peoples, until they gave way to the irresistible influence of civilisation or to the destroying sword. And we are induced to admit the said chronological limits of the stay of the Boii in our plains,* when we observe the considerable quantity and measurement of the *marl earth* in position, the origin of which we cannot but ascribe to this nation, and the heaping up of which would require several centuries. If the Galli were not the inhabitants of our *marl-beds*, I may further add, we find evidences, which we describe in our memoir, which show that they had, in common with that nation, their faith, their usages, their arms and their implements.

With this we bring to an end our discourse, such as it is.

---

which he founded it was "a home-sprung Gallic one, however he became acquainted with it, and, as such, is worth recording." The geographical part of it, he and Arnold, following him, believe to be correct, but they both think the time of their residence in Umbria, before crossing the Apennines, must have been considerably shorter than 200 years; and the latter speaks of the *improbability* of their remaining north of the Apennines, which was contrary to the impulsive character of such barbarians; but there is nothing in the arguments they advance which would prove it impossible that certain tribes, in comparatively small numbers, made settlements in the marshy places and forests, leaving the high lands still to the Etruscans and Umbrians, who would have to be attacked in force to enable these people to cross towards Rome. The fact, by Niebuhr accepted, that those who came last went furthest, would seem to prove settled appropriation of the land in the first coming, as it was the fact of the land being occupied that sent on the new hordes to more southern and western conquests. *Vide* Nieb. *Rome*, ii, 517; Arnold, i, 518.—EDITOR.

* Kohen, Schweigh, Contzen, following Strabo, are of the opinion that the Boii, after their defeats experienced in the years 563 to 571 A.U.C. (191 to 183 B.C.), were *entirely* expelled from Italy. Admitting this assertion, on the one hand, and that of Lopez, on the other, it would seem to establish that the Boii were not in our country more than 213 years, from 396 to 183 before our era.

You must accept it with kindness, in consideration of our love for science and our goodwill.*

Parma, Jan. 1862.

PELLEGRINO STROBEL.
LUIGI PIGORINI.

The excellent results in regard to the study of the ancient races which have been obtained in Switzerland, Germany, and

---

* It may be well here to allude to Niebuhr's idea, that the Etruscans had their origin in the Rhætian Alps, some remains of their language still lingering in the Tyrol. They, according to him, extended southwards into Tuscany, separating and driving back the Umbri, or Umbrici, and the Ligurians, the two most ancient races, whose special habitation was the very Valle Padana in question in this memoir. It is clear that the Etruscans founded towns very early, and indeed their remains would have undoubted evidence of their origin about them. The Umbrians were an Italian nation of great antiquity, who at one time lived considerably further to the north than in later times, when they were driven by the Gauls to the mountainous country south of Rimini, on the left bank of the Tiber. They had also some scattered towns on the coast, says Niebuhr, i, 144, and near the Po, preserved by the lagoons, and partly by paying tribute to the Gauls. It is probable that as their towns are mentioned that their remains could not be confounded with the Celtic pile dwellings, which were, in all probability, the earliest settlements of their Celtic conquerors.

The Ligurians, a very ancient people akin to the Iberians, were the inhabitants of Piedmont before the Gauls made their appearance there; "the whole of Piedmont, to its present extent," says Niebuhr (i, 164), "was inhabited by the Ligurians; Pavia, under the name of Ticinum, was founded by a Ligurian tribe, the Lævians." (Pliny, ii, 21.) They dwelt as far as the Pyrenees, and were driven to the south coast of France by the Celts, and became mixed with them; whence Strabo speaks of the Celto-Ligurians there. Niebuhr says, "Which of the tribes, among the Lower Alps, were Ligurians, and whether the Vocontians were so, I have no means of determining; but from these traces it seems to me extremely probable that this people was dwelling of yore from the Pyrenees to the Tiber, with the Cevennes, and the Helvetian Alps for its northern boundary. Of their place in the family of nations we are ignorant; we only know that they were neither Iberians nor Celts. Dionysius says, "their extraction is unknown" (i, 10). Cato seems to have made inquiries amongst them, but to have heard nothing beyond stories, which were evidently groundless, and clumsily fabricated, whence he stigmatised them as illiterate, lying, and deceitful; and probably they were, having to eke out life with such hard toil, and being unable even to till their stony ground with the plough. The rest of Cato's odious picture is by no means confirmed by other ancient writers," p. 165. Further on Niebuhr says, "A difficulty, indeed, seems to arise from our finding that Polybius remarks that the Venetians differed but little in

Denmark, from the division of the human race into the *age of stone*, the *age of bronze*, and the *age of iron*, should, in my opinion, make us introduce the same divisions also among ourselves, who are newly come to this kind of observation and research; the more so, as with the advance of discoveries we see ever increasing the points of similarity in the productions of our ancient races, and of those who inhabited countries more to the north.

I have premised this that I might repeat, now that the report of Sig. Strobel and Pigorini has given us much additional information regarding the *marl-beds*, that the greater part of the objects discovered in these curious deposits belong to the *age of bronze*.

In fact the number of objects in iron, found in the *marl-beds*, are most limited in number; and we observe that the nails of such metal, of which the authors speak (in p. 35), close up open holes in vases of potstone (*pietra ollare*) *worked upon the lathe*, in view of which we should be obliged to suppose that the lathe was known to the ancient people of the marl-beds, because it must have been used in the construction of the earthenware and some of the utensils in bone. We note that the lance-heads in iron, of which the same authors speak, are probably of the middle age; there is no doubt regarding a knife and some scoriæ; but we must not forget that all the *marl-beds* at present scientifically examined, are deposits which have been disturbed and rearranged (*rimaneggiati*), and that consequently in some of them are found objects of the Roman and of later epochs. Very large, on the contrary, is the quantity of arms, instruments, utensils, and ornaments

---

customs and dress from the Celts, though he tells us their language was not Celtic, he does say it was Illyrian." This allusion to the identity of customs in the Venetii with the Gauls is more singular as they at once sought Roman protection from the Gauls, and were in many respects very different from them and the Ligurians; their country was very limited. However, it is important to consider whether there may not have been a greater similarity and manners in the very early tribes north and south of the Alps, even when not allied by their origin, than the memoir would seem to allow.
—Editor.

in bronze, which are brought to light in the *marl-beds*, and, as we shall see in the sequel, in the *turbaries*.

If we rarely meet, in our public collections of antiquities, with objects in stone,* because our archæologists, up to the present time, have occupied themselves almost exclusively with classical antiquities, all are more or less richly furnished with pre-Roman utensils and instruments in bronze. Our own museum contains a fine collection of *paalstabs* and celts, and other objects in bronze, but unhappily, as is usually the case, wanting any indication of their derivation; we may, however suppose that they have in part come from the valley of the Riparia.

All these circumstances would authorise us to believe that the *age* distinguished by the name of *bronze* had a long duration in Italy, and that our country, especially the hydrographical (*idrografico*) basin of the Po,† was in that age numerously peopled. In stating such a proposition, we bear in mind that the *marl-beds* occupy, at the foot of the Apennine, a zone of considerable size, which extends, almost without interruption, from Piacenza to near Bologna. We should not be surprised, should ulterior discoveries and studies oblige us to attribute to that *age* the instruments of stone found in the environs of Imola and Ancona; the more so, because hammers, hatchets, and lanceheads in stone, like those of Imola and Ancona, were found beside the skeletons of Cumarola associated with arms in bronze; so that, strictly speaking, we should only be able, in Italy, to refer to the *age of stone* the arms found in the neighbourhood of Spezia, in the caves near Leghorn, Mentone, and in Sicily. If the people of the *marl-beds* had possessed or been able to use iron, they would, probably, have fabricated chisels and knives, like those in

---

* I owe it to the courtesy of Cav. Orcasti to have been able to give in Table VI, p. 14, the figure of a knife of flint, which exists in our museum, among the objects of antiquity which have come from Egypt, and which is the only instrument in stone in that collection.

† I learn from Sig. Demortillet and M. Desor, that on the shores of the Lago di Garda there have been lately found, near Peschiera, many utensils and instruments in bronze.

bronze, and they would certainly have used it for points of awls; but all these utensils are of bronze.

I have no need to mention here what a rich harvest we may hope for from the *marl-beds*, especially should there be undertaken regular excavations in the pile systems, the only deposits in position now known to us.

It is probable that the remains of the stations of the ancient races which, more or less altered and reformed, compose the *marl-beds*, are of later date than the piles; we, in fact, see that at Conventino di Castione the *marl earth* rests upon the top of the piles (fig. 5, p. 20). Besides, if the people who gave origin to the *marl-beds* lived upon the piles planted in the marshes of the Po and on the banks of the said marshes, we should find, it seems to me, in the *marl-beds* a greater quantity of the spoils of aquatic animals, molluscs, fish, birds, etc.

We have, then, three orders of stations,—the *caverns*, the *pile dwellings*, and the *marl-beds*.

It is not beside the question to record further, that the arrowheads of flint with which I was favoured by the engineer De Bosis, were found at Monte Oro, near Castelfidardo, and on the hills of Barcaglione.

As for the arrowheads, the hatchets, and the hammers, described by Scarabelli, they were found on the hills of the Imolese; and as for the two arrowheads in flint, which we owe to the courtesy of archpriest Grassi, they came from the hills in the neighbourhood of Sassuolo.

These circumstances show that there were stations also on the hills, forming the outliers of the Apennine. The figure (fig. 27) below represents, in profile, a hatchet, probably of saussurite, found some years since upon the Apennine, forty miles from Piacenza (the notice and profile were courteously given to me by Count Pallastrelli); and further on I shall speak of a similar one found in the Ligurian Apennine.

At points lower down, when the population was more crowded, their products, their refuse, their remains, acumulated in greater quantity, and consequently these remains and their refuse washed and exposed by the rains, rearranged and transported by the water of the torrents formed, perhaps, the *marl-beds.*

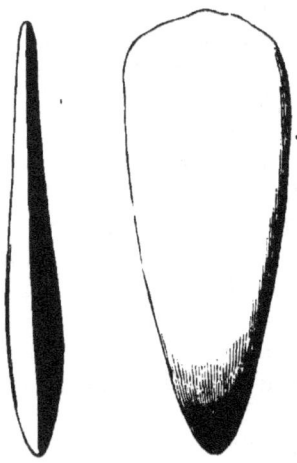

Fig. 27.

I will not hazard, at present, other conjectures, since our researches upon these deposits date from too short a period; and although they have been undertaken with the highest care and learning by the authors of the preceding report, still much remains to be done to clear up their formation, and to know better and appreciate the objects of industry which they contain, and the age to which they belong.

We must still remark on some discoveries made in the *marl-beds,* or neighbouring regions. I have received in the course of the year, from Professor Döderlein, two skulls of men, well preserved, come from the marl-bed of Torre della Maina; they will be described by Professor Nicolucci. I have received from archpriest Grassi, of Sassuolo, a plate of terra cotta, pierced by ten holes, which appears to have served as a gridiron; it was found in the *marl-bed* of Massini, a dagger with handle, and a little knife-axe (*coltello ascia, couteau-hache*) of bronze, as well as a mould of potstone, which seems to have served for founding daggers, such as those alluded to,—these last objects came from the *marl-bed* of Castelnuovo (Reggio).

Finally, Professor Capellini announces to me that he has found among the objects left by the late lamented Professor Alessandrini, an arrowhead in flint, discovered in the environs of Castel Bolognese.

In finishing here the part of this work dedicated to the *marl-beds*, I must relieve my conscience by saying that, looking at the importance of the observations so kindly communicated to me by Signori Strobel and Pigorini, I entreated them to allow me to associate their names with mine on the title-page of this work, but in vain.

SARDINIA.* Those curious perforated discs in terra cotta, of very various form, which Cavedoni calls spindle-whirls (*fusajuole*), are found not only in the country about Imola, Bologna, Modena, in Parma, Piedmont, and Lombardy, but also in the neighbourhood of Cagliari, in the middle of heaps of marine shells, which are met with at some meters above the sea-level. In these heaps besides, as elsewhere, in one of those constructions called *Noraghe* have been discovered fragments of earthenware of the same nature as those found in the *marl-beds;* and, as we shall see in the sequel, also in the turbaries,† nor, as far a sappears, are paalstaves and axe-knives, etc., scarce in Sardinia.

A hatchet of Lydian quartz and three spindle-whirls come from that island, and were courteously presented to me by Sig. Cav. Promis, librarian to his majesty.

PIEDMONT. By reason of the fineness of its workmanship, I should refer to the *age of bronze* the hatchet of *saussurite* given at page 40, fig. 19; it was found in the Ligurian Apennine, in a ravine, of the territory of Belforte, district of Ovada. I owe it to the kindness of Sig. Briata.

Some vases also appear to belong to the *age of bronze*, which

---

* The fact of the occupation of Corsica and Sardinia by the Ligurians, a people of whom Niebuhr says "that they, as well as the Venetians, were unconnected with the history of Italy, and though they dwelt to the south of the Alps, they did so as a branch of nations widely diffused beyond the borders of Italy, while in very early times they seem to have touched in the plain of the Po," would seem to lead us to them as the people who left many of these relics.—EDITOR.

† See La Marmora, *Voyage en Sardaigne*, troisième partie, tom. i, p. 375 to 380.

were found at no great depth, in ploughing up forest land in a district called the *Pennino*,—a hill formed by a moraine, which extends from the heights of Mercurago to Borgo Ticino. These vases, or pipkins, are formed of that same black or reddish paste containing grains of quartz, which are found in the *marl-beds*, and contain generally carbonised bones, and sometimes objects in bronze; two of these, a bracelet and a fibula, I have received as a present from Professor Moro, of Arona, who had himself found them in one of these pipkins. I have learnt, also, that some labourers, some years since, found in one group eighteen similar vases, and each vase was resting on, and covered by, a flat plate of stone. Although of more finished workmanship and more *recherché* form it is probable that certain vases also belong to the same age, or perhaps to the earliest part of the succeeding one (*of iron*),* kindly presented to me by Priest Ambrogio Radice, rector (*prevosto*) of Sesto Calende; these were found in the surrounding regions of S. Anna and of the Groppetti.

I shall here notice a curious vase, in my opinion Celtic,

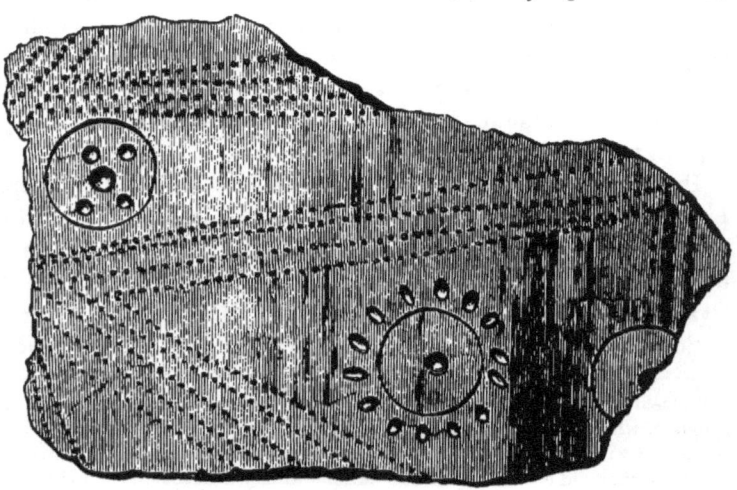

Fig. 28.

---

* These vases are identical with those figured by Giani (*Battaglia del Ticino fra Scipione e Annibale*), fig. 16, 17, and 18, tab. iv; and fig. 4, tab. vii.

discovered in the territory of Bra, in the district of Alba, and of which, unfortunately, only a few fragments were preserved. Its form must have been spheroidal; the substance of its wide base, rising in the centre, forming a pyramidal cone, which enters the internal part of the vase, leaving without a corresponding hollow. The bottom of the vase is occupied by a mass of black clay, which, united to resinous substances with which it is impregnated, must have been exposed to the action of heat, because by its fusion it has been softened so as exactly to take the form of the bottom; a part, also, of the resinous substances, having made its way through the fissures of the vase, still remains, smeared on the outside of it.

I received this vase from my friend, Ettore Craveri, together with other earthenware, with glass tear-bottles, and a set of little discs of amber,* pierced in the centre; these and other earthenware vessels and tear-bottles belonging certainly to the Roman epoch. I am inclined to believe that the amber discs were placed in the vase above described; and it leads me to recollect that I have found, in a similar manner, many of these discs, in part, already altered by heat,—in part, already in a half-fused state; and having experimented with regard to the dark resinous substance adhering to the inside and outside of the vase, it presented the true characteristics of amber. This vase has externally a curious system of ornaments scratched, or rather pricked, upon it, which are not different in character from those used by the Celts.

Morainic Turbaries.

One of the natural beauties of the fertile country which extends from the left bank of the Po to the foot of the

---

* This mention of the use of amber reminds me that amber is found in great quantities among the early remains in Jutland, and seems to have been used as an ornament. Strings of pieces of amber, apparently used as necklaces, as even to the present day, by the Bückeborg peasants in Prussia, having been found; in one case, as I was informed by Professor Thomsen, what appeared to be the stock of a travelling merchant, was found in a morass. It is possible that those mentioned the text in were destroyed at the death of the owner, or partially so by fire; however, it may only have been melted to make a kind of pitch, to be used in some ordinary operation of life. —Editor.

Alps, is its lakes. We may say that all these lakes, beginning with those of Avigliana and Trana to that of Garda, are morainic lakes; that is, contained in amphitheatres surrounded by ancient moraines. The number of these lakes is considerable; some of them are very small, others cover a very considerable extent of ground. When, however, one has given a little study to the regions in which these lakes, as it were, form groups, as in the neighbourhood of Avigliana, those of Ivrea, Arona, Como, one acquires the certainty that, in an age not so very far removed from our time, the number of little lakes and extensions of those which still exist was much greater than at present, and that the small lakes have disappeared; while those tracts of basin, over which the waters of the lakes, still extensive, once lay, are in general become turbaries.

It is in this way that in Piedmont, for example, all the turbaries of any importance, now cultivated, are morainic turbaries. They are found at two different levels, and may be divided into turbaries of the first order, and turbaries of the second order.

When the bottom of the amphitheatre, circumscribed by the moraine, is occupied by a great lake (lake of Orta, Maggiore, or Garda), there are found turbaries whose level is a little above such lake; when the bottom of the amphitheatre is not occupied by a great lake (amphitheatres of Rivoli and Ivrea), but is traversed by a torrent, turbaries are found raised only a few meters above the level of the torrent. These are the turbaries which I shall call turbaries of the first order. They are for the most part very extensive; for example, that of Angera, which is an ancient branch of the Lago Maggiore, and that of Avigliana, which is in continuation of the lake which bears that name.

The turbaries which I shall call of the second order are found in basins rather restricted in size, and placed at a much higher level, on the top,—that is, of the moraine itself. Such are the turbaries of Alice, Meugliano, and San Martino near Ivrea; and those of Mercurago, Oleggio Castello, and Borgo Ticino near Arona.

It is generally in the turbaries that, in Lombardy and Piedmont, are discovered ancient objects of industry, and especially of the *age of bronze*.

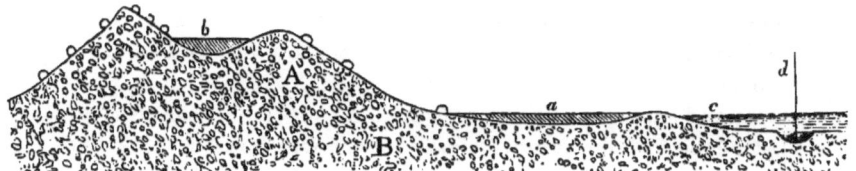

Fig. 29. a. *Turbary of the 1st order.* b. *Turbary of the 2nd order.* c. *Lake occupying the lowest part of the amphitheatre. When there is no lake, the bottom is traversed by the torrent* d. A. *Erratic part of moraine.* B. *Diluvium.*

In the year 1856, two brothers, Signori Villa, of Milan, published a short notice, illustrated with two good figures of two hatchets (*paalstab*) of bronze, and sundry flints cut into arrowheads, found in the turbary of Bosisio, at a depth of about three meters.

The Signori Villa give as their opinion, that these worked-up flints are "of the nature of those which are found in the *majolica* marble, lying over the red mountain limestone (*calcaria rossa ammontifera*) of the mountains between the Lago Maggiore and that of Garda." This opinion is not, however, shared by any of the geologists of Lombardy that I have had occasion to consult on the question; and certainly the fragmentary state in which the flint (*selce piromaca*) exists in the specimens with which I have been favoured by the engineer De Mortillet, leave grave doubts on my mind whether, with such a variety, it would be possible to obtain arrowheads. Professor Döderlein has, however, assured me that, at Enego, in the valley of Astico (in the territory of Vicenza), some years back, flints were extracted from the *majolica* and worked up for guns.

Objects in bronze were elsewhere discovered in the turbaries of Brenna, lying near skeletons, which unfortunately, being suddenly transported and buried in a neighbouring cemetery, were withdrawn from possibility of examination. And finally, Sig. de Mortillet tells me that he has seen at Verona, in the possession of Sig. Martinati, a hatchet of bronze, coming from the turbaries of the Venetian territory.

Turbaries of Borgo Ticino and Mercurago, near Arona. The most striking discoveries have been made in the turbary of Mercurago, a little place, half an hour's walk from Arona. There Professor Moro, besides arms and implements in stone and bronze, and—in addition to utensils, tools, and vases in stone—wood and terra-cotta, discovered a pile system in position, and such conditions as would persuade one that, in the little lake of Mercurago, before the turf was formed, lacustrine dwellings existed of the same kind as those, the remains of which are found in nearly all the Swiss lakes.*

Having made some visits to the turbary of Mercurago, and having besides in my possession the objects found, I am able to give here a minute account of them. There had been found at various times, in the course of years, in digging out the turf, arrowheads of flint, and the head of a javelin of bronze, as well as many fragments of pottery made of blackish clay, embedding grains of quartz; and finally, a wooden anchor, more than a meter in length, which terminated at one end in two hooks, and at the other, was perforated to receive the rope. In this year (1860), at a little distance from the place where the anchor was found, first a canoe, formed of the trunk of a tree, hollowed out, 1·90 m. in length, about a meter

Fig. 30.

in breadth, and 0·30 m. in depth. In spite of the care which, in consequence of the warm recommendations of Professor Moro, was taken in extracting and preserving it entire, this canoe was not slow in breaking up, separating—owing to the long maceration of its sides,—and dividing into longitudinal strips as it dried. When I saw it, you could still trace on its

---

* See, for these, the memorials of M. Keller in the acts of the Society of Antiquaries at Zurich, and that recently published by M. Morlot, "Études Géologico-Archéologiques," in the *Bulletin de la Société Vaudoise des Sciences Naturelles*, and the *Habitations Lacustres* of Sig. Troyon, already quoted.

bottom the marks of the instruments used to excavate it. Not far from the canoe was also found a drill of bronze, and a disc of terra-cotta, with a hole in the middle, like the spindle-whirls found in the *marl-beds* of the Modenese, and near Imola.

The form of the turbary of Mercurago is oblong, and all the objects, of which I have made mention, were found in a narrow space at the northern extremity, about forty meters from the edge, and in a position in which the water (when the turbary was a lake) could have had, at most, two or three meters of depth. In this same place, in opening a ditch, was discovered a series of piles, in diameter from 0·15 m. to 0·25 m., fixed vertically in the grey mud which forms the bottom under the turf.

Having been courteously informed by Professor Moro of this discovery, I repaired with him to Mercurago, when the intelligent director of the works of the turbary, Sig. Maffei, had extracted, with much care, one of these piles, and we were able to persuade ourselves that the instrument used in sharpening it must have had its cutting part in a curved shape, as the traces it has left are sensibly concave.

The length of the piles is from 1·60 m. to 2 m.; they are planted deeply in the mud, and are elevated a meter of their

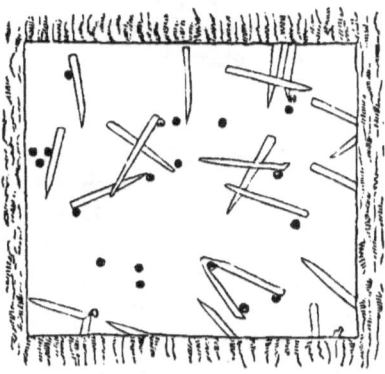

Fig. 31.

height in the mass of the turf, which further covers them with a meter of substance. At the present time, the superficies of the excavation is square, its sides measuring about nine meters. On this superficies were found twenty-two piles, planted vertically, and bound together by some cross pieces. At the bottom of the stratum of turf, and in the plane of separation between the turf and the mud, was found among the piles a bed or litter of crushed flints, upon which were found enough

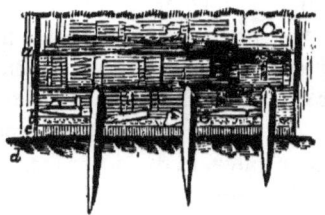

Fig. 32.

fragments of pottery to fill a large basket; three earthen vases in good preservation, a large needle (*spillone*) of bronze, a lamina (*scaglia*), some arrowheads, and many small laminæ and fragments of flints, hazel-nuts (*nocciuole*), cornels (*corniole*), etc.

All these objects, and the conditions under which they were found, demonstrate that on this system of piles, planted in the lake at a little distance from the shore, was constructed a habitation in which lived, before the turf began to be formed, a human family; and the fact of finding, together with worked flints and red clay, earthenware with grains of quartz embedded, a needle of bronze, shows, besides, that that family lived in the time of transition from that of *stone* to that of bronze.

The arrowheads are generally of highly finished workmanship; the lamina of flint, on which I have remarked, is 0·122 m. in length and 0·017 m. in breadth; longitudinally it is curved in the shape of a bow, and has the lower superficies formed of a single face and smooth, and the upper, of two oblique planes which cut longitudinally a third face; the two oblique planes, of which the upper superficies is composed, coming to meet the plane forming the inferior superficies, which results in very sharp lateral angles, and in other words, two edges; one of the edges has been reduced to the form of a rude saw by means of a series of blows of a small hammer, which, by detaching equal flakes, produced a kind of serrating. What is remarkable in this lamina (which we may consider as a rude knife-saw) is that in its general form it is exactly analogous to certain laminæ of obsidian, which the cabinet of mineralogy of the School of Practical Engineering received from Mexico, together with some specimens of the same rock, having the form of an olive very much elongated,

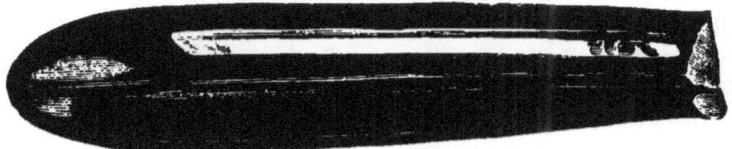

Fig. 33.

and which present wide longitudinal furrows. The form of the laminæ would suggest that they were split from similar masses by a smart stroke of a hammer upon one extremity. The absolute identity of form between the laminæ of obsidian from Mexico and that of fire-giving flint (*piromaca*) found in the turbary of Mercurago, would make us presume that this last was split, with the same process, from an oblong mass of flint.*

However, as I proved to myself, notwithstanding my habitual experience in breaking stones with a hammer, that I could not succeed in splitting, from these masses of obsidian, fragments presenting any resemblance with the laminæ or splinters in question. I read with much pleasure in the letter, full of talent, published lately by Sig. Delanoüe (*De l'Ancienneté de l'Espèce Humaine*), an account of the true process used by

---

* See Troyon, *op. cit.*, p. 462, and Table v, fig. 22. My friend, Sig. Craveri, who lived for sixteen years in Mexico, having examined these masses of obsidian with the lengthened olive-shape, assured me that these are really the remains of the masses from which the natives have split off the splinters, which used to serve them for the manufacture of cutting instruments; and added, that a large quantity of such masses of splinters of obsidian are found at Cerro di las Navajas (the hill of the knives), near the cordillera of Pachuca.

the Mexicans to obtain their knives of obsidian.* Signor Delanoüe, speaking of the hatchets of stone which, by hundreds, are found in the ancient alluvial deposits of Amiens, informs us that the workmen of the place succeed in counterfeiting them; and adds, "The true ones bear the traces of long and thin (*minces*) undulating rays, such as our workmen cannot produce by the hammer. The manner of making them (that is, the one used by the aborigines in making their hatchets) would be analogous to that which the Spaniards found in use in Mexico for the manufacture of knives or razors in stone. *The Aztecs detached pieces of obsidian, not by a blow, but by a particular sort of pressure.*† These knife-razors are exactly like those long narrow knives of flint of the caverns, which our ordinary processes could not certainly reproduce at the present day."

These and many other analogies of form and construction, which are noted between the instruments and utensils of the ancient races which peopled our lakes and marshes in the age of stone and bronze, and those of South America and the islands of the Pacific, show that wherever they may be and in all times, man, pressed by the same wants, has used nearly the same means to succeed in satisfying them.

Generally, the earthenware vessels are of very rude workmanship, and made without the help of the wheel. The two vases given below (figs. 34, 35), were found still furnished with the cord which, tied to the two handles, served

---

* The Apaches, and various other wild tribes of Northern Mexico, using arrows still in the present day exclusively in the chase. Some of these arrows are of obsidian, but the greater part are of flint; not because obsidian is rare in those regions, for it is very abundant all through the republic, but because the flint is preferred, as far as appears, for its greater power of resistance.

† This process (pressure) is in use to the present day among the Indians of Mexico in making their arrows. Sig. Craveri, to whom I have referred above, has told me, that when they wish to make an arrow, or other instrument, of a splinter of obsidian, they take the piece in the left hand, and hold grasped in the other a small goat's horn; they set the piece of stone upon the horn, and dexterously pressing it against the point of it while they give the horn a gentle movement from right to left and up and down, they disengage from it frequent chips, and in this way obtain the desired form. Compare Tylor's *Anahuac* for a more minute description.—EDITOR.

to hold them suspended; it is formed simply of twisted oziers.

Fig. 34.

Fig. 35.

The vases found at Mercurago, if they are compared with those of the bronze age, which I have received from Switzerland, turned out in the comparison much more rude, not only in the workmanship, but also in the paste of which they are composed; the walls are much thicker, and so in proportion the vases heavier. There are found, however, some vases with ornaments, but very simple, executed by scratching or pricking while the paste was still soft, and consisting in triangular figures drawn with lines parallel to the two sides.

Those which were presented to me by Sig. Radice approach

in form, kind of workmanship, and thinness of their sides, those which I possess from Switzerland; these, also, have ornaments etched on them, consisting of lines sometimes running horizontally parallel, but more or less distant from one another round the vase, sometimes in series of short oblique lines, from which association arise geometrical figures (like those remarked as on the fragments found at Mercurago), or rude imitations of branches furnished with leaves.

This kind of ornament gives to the vases found at Sesto Calende a certain degree of likeness to some of those found in the tombs discovered, and illustrated with such loving care by the Marchese G. Gozzadine.*

Among the Roman earthenware of various form and workmanship which I have had occasion to examine, few are to be found with ornaments in etching (*scalfitura*); however, there is an instance in the excavations made at Turin in opening the Cernaja road, and particularly in laying the foundations of the embankment of the railway, Victor Emanuel; these were found near the ruins of the old walls, the remains of a vast cemetery. The tombs consisted, for the most part, of large amphoræ, terminating inferiorly in a point, and above in a narrow neck, and having their bowls sometimes oblong, sometimes spheroidal. In these amphoræ were found lamps, flasks, statuettes; various vases in red earth, some of them of high finish, and with ornaments stamped or worked on them; glasses of various form and colour; imperial medals; objects in bronze, in iron, in ivory, circular and square; plates of a metallic alloy, very hard and very highly polished; bones in fragments, etc., etc. Since the dimensions of these objects would not have permitted them to pass down the narrow necks of the amphoræ, they must have been cut transversely below the neck, where the bowl becomes larger, and the objects

---

* *Di un sepolcreto etrusco scoperto presso Bologna.* The author has already remarked the analogy which exists between some of the vases, the figures of which he has given, and some of them illustrated by Giani (*op. cit.*), which came also from the environs of Sesto Calende. I am most grateful to Sig. Cav. O. Fabretti, for having made me acquainted with, and communicating to me, the elegant work of Gozzadini.

being placed in the lower division, it was covered with the upper, and the mouth corked up with a disc of terra-cotta or a plate of bronze. Among all these amphoræ,—from all the various kinds of pottery, which as was clearly seen were in fact new,—one small vase was found of terra-cotta, whose black spots showed that it had been exposed to the flames. The smaller vases presented a rough ornament in etching, the lines sometimes cutting each other alternately, sometimes not, obliquely to the axis of the vase.

Although I here feel the necessity of declaring myself absolutely uninitiated in fact in archæology, still, I think I may say that the Etruscan cemetery described by Gozzadini, to which I have alluded above, is of posterior date to the cemetery discovered in 1856 at Cumarola, near Modena, and later also than the age in which were manufactured the objects which, at the present day, we discover in some of the *marl-beds*.

It is true that the arms found with the skeletons at Cumarola can hardly be compared with the utensils discovered in the Etruscan cemetery of Villanova, near Bologna, because from that there were only taken arms in very small number; but however this may be, one feels that in an epoch in which men knew how to unite with bronze iron and enamel (*smalto*), in an age in which they made porcelain in such a masterly way, they would no longer have worked at stone-arms, nor constructed those rough vases the fragments of which are found in many of the *marl-beds*.

If, therefore, it appears evident that the tombs described by the Marchese Gozzadini are later than the date of burial of the skeletons discovered at Cumarola; by analogous reasoning they are later, also, than the ancient lacustrine habitations of Mercurago. It is no less true that there exist, between all these remains of remote epochs, some points of contact. In fact, ornaments similar to those which are seen upon fragments of pottery found at Mercurago, are reproduced, as we have already said, upon vases coming from the neighbourhood of Sesto Calende, reproduced also on some of those from the Etruscan burial-ground of Villanova, near Bologna. If, then, we examine the beautiful plates published by Gozzadini, we shall perceive that, abstracting the work of highest finish, the *spindle-whirls*, given by him in Table VII, have almost exactly the same form as those found at Imola, in the Modenese, at Sesto Calende, and generally in the palisades of the lake of Mercurago, or of the Swiss lakes. We shall perceive, further, that some of the needles and knives are in form identical with those found at Nidau, Estavayer, and Chevroux, in Switzerland; that all the ornaments of the pottery, without distinction, are executed by etching; and we could adduce many other points of connexion, if this work were not, for good reasons, intended to give notes rather than dissertations.

The javelin head (bronze) found in the turbary of Mercurago and those found at Cumarola, near Modena, have a different form; but all present the peculiarity of being very thin, and having on the lower side a part jutting out, like a small tail, which was intended to enter and be grafted into a cleft made in the extremity of the handle, and further strengthened by one or two rivets; the thinness of the blades and the mode of insertion show how little solid and formidable these arms were. I have seen in some of our collections other lanceheads in bronze; but probably they were of a more recent epoch, as they are already furnished with a fit sheath for the insertion of the handle. The same mode of insertion, that is, by means of a prolongation of the lamina, fastened to the handle with rivets, is observable in the bronze sword discovered in the marl-bed of Marano, and described by Cavedoni, and in some lanceheads coming from Sardinia (Royal Armory).

Lately, discovery was made at Mercurago of a utensil or machine of wood, of very curious workmanship. It has the form of a wheel, not however exactly circular; in the middle it has a hole, into which enters a piece of tubing, after the fashion of a nave; and between the nave and the periphery are two empty spaces in the form of crescents. Three pieces of wood (probably walnut-wood) together compose this machine; and to keep them united, they are clamped together by two strengtheners about the middle (*a metà legno*), bent in the shape of a swallow-tail; the strengthener does not run in a right line, that is, parallel to the diameter of the wheel, but is bent nearly parallel to the periphery, so that to make it enter the space, it would be necessary to make it pliable; the

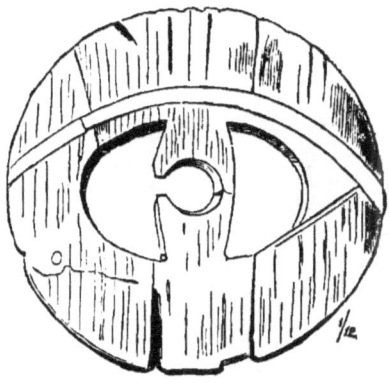

Fig. 36.

strengthening piece is of larch-wood, and carbonised on its lower side.

Professor Moro being aware, from experience, that the objects in wood found in the turbaries cannot be preserved, because on drying up they completely lose their form, sent the utensil in question to me incased, and wrapped up in mould and hay saturated with water, so that it arrived at Turin in the soft (*pastoso*) state in which it was found, and Sig. Comba was enabled to take a model of it in plaster, which fully corresponds with the original, which, as is usually the case, became reduced to minute fragments.

At first, I remained in doubt whether the utensil described could have served as a wheel; afterwards all doubt was removed by a new utensil or instrument analogous to it, but of much more accurate workmanship, coming in the course of the present year from the same great bed; it is a wheel of elegant form, in which, as in the preceding one, there is not the slightest trace of any metal. While Sig. Comba was in the act of taking its copy in plaster, he observed that where the fibre of the wood came at the part where the machine, when in action, touched the ground, grains of sand were embedded in the mass of the wood; and on examining the cor-

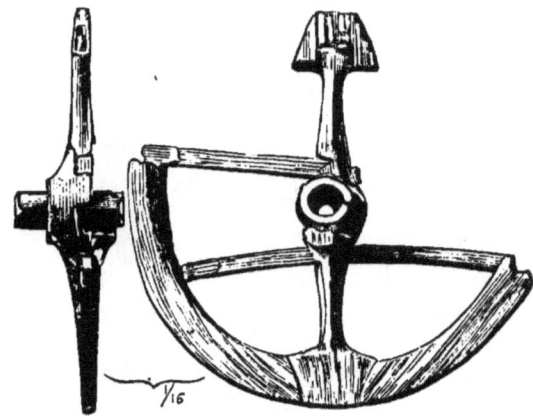

Fig. 37.

responding pieces of the first wheel, the same fact was repeated,—that is, the presence of grains of sand embedded superficially in the mass of the wood. It seems thus sufficiently demonstrated that these two objects are wheels, and served to such use.

In the turbary of Borgo Ticino (a turbary of the first order), the position of which is nearly on the level of the Lago Maggiore, are found vases of earth, just like those discovered in the turbary of Mercurago. There were found also a great number of earthen vases, arrowheads in flint, and objects in bronze, in the neighbouring turbary of Gagnago; but all have been broken, sold, or dispersed. On the bank of this basin, I have also seen fixed in the turf piles identical with those of Mercurago, but which present the peculiarity of bearing traces on their upper extremities of having suffered the action of fire; I may add, that charcoal and ashes, and particularly boughs and trunks of trees, with traces of fire on them, are frequently met with in the neighbourhood of Arona and Ivrea. The trunks of trees, which are found in abundance in these turbaries, belong for the most part to the tribes of pine, oak, alder, birch, willow, and walnut, etc., and appear generally to have lived on the spot. Twenty years since the turbary of Gagnago was a pasture-ground; five to six meters of turf have been extracted, and now the basin has become again what it was anciently,—a lake.

In the neighbouring turbary of Conturbia have been discovered objects of a more recent epoch, among which I shall mention some piles, fixed nearly in the centre of the turbary, and I have been informed by the agent that one of these piles was furnished at the bottom with an iron point. I have been able to procure one, deprived however of its iron point; and I saw that it had been sharpened with the same sort of instrument which we should use for the purpose.

The turbaries of Mercurago, Gagnago, and Conturbia, are turbaries of the second order; the same is the case with that of Revislate, leaving which to go to Borgo Ticino, one meets half way an enormous erratic mass, called in the country *Pietra Grezzana*, this measures one way fifteen meters, another ten, and rises four or five meters from the ground.

Turbaries of San Martino and Torre Bairo near Ivrea.— The results obtained from the researches initiated in the turbary of Mercurago by Professor Moro, encouraged me to visit those of San Martino, situated near Ivrea,—a visit which I undertook with so much the greater pleasure, inasmuch as having been in correspondence for many years with the excellent Doctor Gatta, administrator of our society, who is proprietor of this turbary, I not only had him as companion in my inspection, but obtained from him all the information and all the means which could make more easy for me the observations which I had conceived the idea of making.

The turbary of San Martino is situated on the back of the moraine, at the spot where—ceasing *at the priests' bridge* to be lateral, to run, that is, in a right line—it begins to bend into the shape of a bow and become frontal; the basin which it includes is nearly oval, its largest diameter measuring nearly two chilometers, and the smaller, one chilometer.

Surrounded by meadows and fields intersected with clumps of trees of high growth, this basin, on the north bank of which,—in the midst of chestnuts and walnuts,—rises the town of San Giovanni, has a gay and picturesque aspect, a quality which one seldom meets with in turfy places.

For some years the labours of extraction have been ably directed by Sig. Barbano, and I have learnt from him that, up to the present time, in that turbary no traces of regular pile systems have been found which would lead us to suppose that lacustrine habitations existed in the ancient lake San Martino. There have, however, in these last years been found two earthen vases and a manufactured flint; one of these vases was courteously presented to me by Doctor Gatta; the other and the flint had been already given away to Sig. A. Sismonda. That which I possess, made without the assistance of the wheel, is of very rough workmanship, but still there is evidence of a desire to ornament it; for at a distance of some centimeters below the brim, it has a circle of stamps roughly made with a splinter of wood-stone, or other hard body. Its form is nearly cylindrical, having only a slight lessening towards the base; the paste, like in all respects to that of those vases found at Mercurago and at many places in Switzerland, is a black clay, with small grains of quartz in it. I have just received from Sig. Barbano a curious spindle-whirl found in the course of the year in the same turbary. The fact of having found a manufactured flint, and the utensils of earth of which I have made mention, proves that in the bank and in the vicinity of the lake which once occupied the turfy basin of San Martino, a Celtic tribe had their residence; and it gives one reason to believe that we shall end by finding in the same basin the remains of lacustrine habitations. Nay more, the discovery of such remains appears to me to be likely—for not having at present excavated more than a first and superficial stratum of turf, and there remaining to be extracted a second, in some places more than a meter in depth—it would be in this lower stratum that we must meet with heads of the piles, if any exist. In this, and in other less turbaries round it, there have further been discovered fish-baskets (*nasse*) of osiers; in one of these turbaries, situated in the district of Torre Bairo, pieces of earthenware were discovered, which seem to have been manufactured with the wheel; and, finally, in another was found a small millstone. And here we must remark that, in giving a value to the different objects of human industry met with in these turbaries, it is well to proceed with caution,* if it be true that in their vicinity they were frequented by populations who lived long before the common era, which appears sufficiently demonstrated from the vases and manufactured flint found in the turbary of San Martino. It is still undoubted that these same regions were very populous in the Roman epoch, which may be deduced from the number of tombs in terra-cotta which one meets with in ploughing up the forest land,—tombs which ordinarily contain pottery of the most varied form.

To make possible and easy the extraction of the turf in the basin of San Martino, it was necessary to open, with considerable but necessary expense, a deep canal to drain it, which cutting into nearly its whole length, permits one to study its soil. Generally speaking, this is formed of a whitish clay, which, when massed, appears regularly stratified, and which elsewhere allows of subdivision into very thin beds, which present the impress of herbaceous plants.

In some places, one finds resting on this clay a considerable stratum of a blackish substance, rich in combustible material, which on drying becomes hard, breaks most irregularly into small pieces, and is reduced, finally, into minute fragments. If this same substance is manipulated, that is, formed into a paste and made into small bricks, it gains a great deal of cohesion, and becomes an excellent combustible; best if in making the paste there is mixed with it a certain quantity of turfy fibre. The substance of which we speak, or to speak more correctly, the blackish mire, extends in some places fifty centimeters, and is a meter or more in thickness, and on this rests the real turf in greater or less thickness. In other places all, or nearly all, the turfy mass is formed of a stratum of

---

* In a turbary, or *moregna*, situated near the lake of Viverone, was found, some years back, a sword of bronze, now in the possession of Sig. Count Mella, of Vercelli.

mud, of a dirty brown colour, tending to blackish, which on drying swells, subdivides, opens into very thin laminæ, like a badly bound book whose back is too small for the number of its leaves.

The stratified clay, and the beds of mud of different kinds which cover it, show clearly that the turfy basin was formerly exclusively aqueous, or in other words, that it was a lake. Finally, the subsoil is composed of erratic deposit, with masses of considerable volume and flintstones.

Starting from the centre and walking towards the southeast side of the basin, it is apparent that the turf diminishes in thickness as we approach the edge of the turbary, and there the spade in cutting the clods is often stopped by encountering great trunks of trees, which may be seen at different depths projecting from the walls, produced by cutting into the turfy mass. Some of these trees are of considerable dimensions, thirty, forty, fifty centimeters in diameter, and the roots are often found with the trunks; so that there is no doubt that they lived on the spot. They are usually pine, oak, nut-trees, and alder, etc. It happens very often that we meet with several together, laid one on top of the other in a pile, they lie in very different positions; but, taking them all together, it appears that they tend preferentially to turn their tops towards the centre of the basin, which does not prevent some lying somewhat parallel to the direction of the bank. Generally they are covered with a meter of turf; but the thickness of this diminishes, as we have said, as we approach the edge, and at the same time the number of trunks increases.

These observations are made on a surface of considerable extent, and compel us to argue that a whole forest was laid low on the banks of the basin before it was occupied by the turf. At the first glance, also, we see that the turf which covers this ancient forest is subdivided, as in the centre of the basin, into small strata of different colour and texture, which all bend themselves, like a bow, over each trunk, so that as we run the eye over a cutting or wall, which runs through a good piece of the turbary, we see that the turfy strata, instead of running parallel to the plain which forms the bed of the turbary, present a series of undulations corresponding to as many trunks or groups of trunks.

If I have properly given a value to the facts above noted, it would appear to result from them that at first there was in the middle of the basin a pool, in the bottom of which was deposited the mud of which I have spoken, and that little by little, in the sequel, the lake and marsh vegetation becoming developed crept in part into the watery mass; and that close to the pool grew a forest of trees which, falling from age or overthrown by some hurricane, were in time covered by the turfy vegetation which grew from the centre towards the edge of the basin. It is known that turfy vegetation has in the highest degree the property of extending itself, and invading the places surrounding that in which it is at first fixed, and particularly of taking possession of old trunks.

In all the other small turbaries, which lie to the west of that of San Martino, the same fact is observable, that is to say, that towards the edge trunks of trees exist in great quantities, buried beneath from one to two meters of turf; and not to forget the question with which we are occupied, I may add that many of these trunks bear the evidence of having suffered burning; but the greatest number of such trunks which I have seen in the most restricted space is in the turbary of Signor Antoniono, situated in the district of Torre Bairo, in the region *Palude lunga*. There are whole trees, for the most part pine and oak, and they are (as I was informed by the proprietor of the turbary, who has, with much discernment, interpreted the facts which have been presented to him) untouched, with the bark,—that is, with their branches, leaves, and fruit (mast and acorns).

Signor Antoniono pointed out to me that, when one comes upon a group of trunks laid one upon another, one observes generally that those below are contused at the point at which they are ridden by the others, and that these at the same point are broken. The quantity of these trunks is such that, from the single cutting followed by him for two years, there have been extracted one hundred and ninety metrical *quintals** of wood; of these trunks, some were found in a good state of preservation, and 15, 16, and 17 meters in length, so that Sig. Antoniono was able to make with them the pillars to support the roof of a cottage of considerable size, that he has had made near the turbary.

In this work I have tried especially to indicate the discoveries recently made in central and northern Italy, of arms and instruments in stone and bronze; and those relating to the existence of ancient habitations in the basins of our morainic lakes for the most part become turbaries, and in the marshes of the Po, now transformed into cultivated and fertile fields.

The unexpected, as it was powerful, assistance which I have received in these studies from Signori Strobel and Pigorini, makes me hope that other naturalists will dedicate themselves to it, choosing for their investigations regions hitherto unexplored in Italy,—investigations which do not present serious difficulties or demand great efforts, but which, nevertheless, may be expected to give results, the importance of which it is impossible to calculate.

Two years ago, I was in fact ignorant that in Switzerland had been discovered remains of lacustrine habitations of the age of *stone*, of *bronze*, and of *iron*. Professor Desor, who had come into Italy to study our lakes from an archæological point of view, initiated me into the research after Celtic antiquities, and in the month of May 1860, wished me to accompany him in a tour which we made to Arona, with the object of discovering remains of pile dwellings along the banks of that part of the Verbano† which, below Angera and Arona, extends as far as Sesto Calende, where the Ticino recovers its course. The indications with which I was favoured by Pro-

---

* A *quintal* is a weight of one hundred pounds, supposed to be derived from the Latin *centum* (Ash, *Dict.*); what a "metrical quintal" is I am not aware, but it is probably analogous to so many "tons measurement" in shipper's language.—EDITOR.

† *Verbano.* This is, I believe, a local name for the lower part of the Lago Maggiore, which, I have observed, is extremely shallow to near Belgirate, if not above it.—EDITOR.

fessor Moro, and those which we collected from some fishermen, led us to expect to succeed in our intention; but the waters at that season always large, dissipated our hopes.

Fortune willed that what we looked for in the lake should be discovered in the neighbouring turbary of Mercurago, and again, a year later, in the marl-bed of Castione, under conditions much more favourable for study. These discoveries are due to Signori Moro and Strobel. The objects found in the turbary of Mercurago were kindly given to me by Prof. Moro; these, although still few in number, are of high interest, and are destined—together with those which I have had presented to me by Sig. Gatti of Modena, by Sig. Grassi, archpriest of Sassuolo, by Sig. Radice, rector of Sesto Calende, and Doctor Gatta of Ivrea—to form the nucleus of a Celtic collection. It is on this account that I do not consider them as mine, but rather as belonging to the public. And turning to those persons who shall wish to occupy themselves, or who shall find themselves in a position to make some discoveries of the same kind, as those to which this memoir refers, I shall end it in the words which M. Vouga addresses to his fellow countrymen, in an article published by him on the "Lacustrine Habitations of Switzerland":—

"The antiquities of a country are its public property; they are historical acts which no individual has any right to monopolise to ornament his chimney-piece, or give as toys to his children. I should like further to believe that the public—those members of it whose attention may be attracted to this article—will only make use of the information it contains in its own interest, namely, by depositing in public collections the objects they may be able to discover."

A collection of objects, obtained from the marl-beds, is being formed at Parma, under the care of Signori Strobel and Pigorini; and I only hope that they may meet, in the proprietors of the marl-beds, with the same generosity and love of science which I found in the proprietors and directors of works of our turbaries.

NOTE.—Signor Collomb, starting from the fact that, at St. Acheul, near Amiens, the manufactured flints are met with in a bed of rolled flints under a deposit of gravel and sand, the whole being covered with *lhem*, or vegetable-bearing earth; and making use, further, of all the elements of comparison, and discussions ministering to them arising out of his profound acquaintance with the *diluvio-erratic* earths, synchronises the stratum of flints of St. Acheul with those which, in the valley of the Rhine, lie under the moraines, and thence draws the consequence, that the origin of man mounts to beyond the glacial period (*De l'homme fossile dans ses rapports avec l'ancienne extension des glaciers.*—Lettre de M. Ed. Collomb à M. Ed. Desor). Although, as it appears to me, there are not serious reasons which oppose themselves to our making the existence of the human race mount back to that epoch; still, I cannot help considering (and in this I fear being too presumptuous), as still liable to discussion, the premises from which Sig. Collomb deduces the consequence I have noted, as it appears to me that the synchronism admitted by him is not fully ascertained.

I consequently adopt the idea expressed by Sig. Desor, in the work in which he attempts to combat the opinions of Collomb (*Des Phases de la Periode Diluvienne, et de l'Apparition de l'Homme sur la terre.* Par E. Desor).

In Italy, the most ancient traces of man are, perhaps, those which have been discovered in the caverns; but there is no proof that the strata in which these traces, remains, or relics of man are found, are either anterior, contemporaneous, or posterior to the grand extensions of the Alpine glaciers. Certainly, as far as my knowledge extends, traces of men have not yet been met with in our *submorainic diluvium*.

I would wish here to state all the discoveries at present made, which militate in favour of the opinion given out by Sig. Collomb.

In the stratum in which is excavated the left bank of the Po, near Carignano, was discovered some years since, a molar of the *Elephas primigenius*. Lately, Cav. Curioni, of Milan, communicated to me a magnificent molar of the same species, together with a cranium of the *urus* which came to light near Chignolo (Pavia), owing to a flood of the Po. It is noted elsewhere that, in the environs of Arena, the skulls of *Urus* and *Cervus megaceros* were found, which adorn the museums of Pavia, Turin, Parma, etc. All these deposits and strata must belong to the same horizon; and it is probable that this horizon may be pre-glacial, or in other words, *diluvial*. It is probable that in these strata we may end by finding, some day, some human remains; but all these probabilities are not yet facts.

I shall here note, that the stations of Mercurago, S. Martino, etc., lie upon the moraines, and are, therefore, far later than them; but these stations belong (certainly that of Mercurago, and probably also that of S. Martino) to the first times of the age of bronze, and prove nothing against the existence of man before the extension of the glaciers.

I have mentioned these stations, because if it is ever put beyond doubt that man existed before the extension of the glaciers, we shall be forced to admit that the age of stone alone lasted thousands—rather tens of thousands—of years, since the glaciers must have needed a no less time to descend to the plain, and construct those gigantic moraines which, to the mouths of the valleys, are met with over the whole perimeter of the Alps, and on the backs of which we now find the stations of the beginning of the age of bronze.

ADDITIONAL NOTES:

BEING

A PRÉCIS OF DISCOVERIES IN THE LAKES AND TURBARIES OF PIEDMONT, LOMBARDY, AND VENETIA IN THE YEARS 1863 AND 1864.

By CAV. BARTOLOMEO GASTALDI.

THE settlements (*stations*, Fr.) of the lakes of Lombardy and Venetia are the most important discoveries made in the last two years (1863-64) in Italy, and, as well as those already made in the turbaries of Piedmont, are due to the initiation of M. Desor, of Neuchâtel. In the months of April and May, 1863, this Swiss savant, accompanied by his fisher Benz, Sig. Stoppani, and M. Mortillet, verified—in the lake of Varese, in that of Como above the bridge of Lecco, and in that of Pusiano— the existence of several lacustrine settlements: some fragments of pottery, of a characteristic paste, sufficed to enable M. Desor to class those pile dwellings among those of the bronze age.*

On the announcement of these discoveries, the Italian Society of the Natural Sciences at Milan, with a view to the encouragement of science, made a grant of five hundred francs for making palæontological researches, and especially for the study of the lacustrine dwellings. The direction of the works was confided to Sig. Stoppani, and no one could have acquitted himself with more zeal and talent. We shall extract from his Report to give an idea of the results obtained.† Keeping solely to the lake of Varese, Sig.

---

* Stoppani, *Prima ricerca di Abitazioni Lacustri nei Laghi di Lombardia*, Atti della Società Italiana di Scienze Naturali, vol. v, p. 154.

† *Ibid.*, p. 423.

Stoppani added new discoveries to those which he had made in the spring, and he was able to make himself sure of the existence of five settlements, or pile habitations; a sixth was afterwards discovered by the Abbé Ranchet. These settlements received the following names :—

Settlement of Isolino.—This is a field of pebbles spread out over the mud, forming the bottom of the lake, and arranged in a circle. The pebbles occupy a space of about four thousand square meters; and from the middle of the pebbles may be seen protruding the heads of the piles, which there is often difficulty in distinguishing. There have been found three splinters of silex, fragments of vases, a great quantity of knives of flint, like those of Mexico, a lance-point, also, in flint; a great quantity of bones worked into piercers and chisels; many teeth, and especially of the stag, the goat, the ox, and the boar, etc.; and finally, a fish-hook in bronze; the flint which has been used generally is blackish, probably extracted from the cretaceous marls in the neighbourhood of the lake; the reddish flints are more rare, and must have been brought from a distance, that is to say, from the Jura limestone.

Settlement of Cazzago, which does not deserve a special account.

Settlement of Bodio.—This is the richest, and it is here that the most remarkable discoveries have been made; the general form of the settlement is circular. There, as well as at Molino, many fragments of vases, teeth, two axes of rock of the nature of serpentine, a great quantity of arrowheads of very delicate workmanship, distinguished especially for the fineness and length of their barbs, some spindle-whirls, and some instruments in bronze.

Settlement of Keller.—This is an exact reproduction of that of Bodio, with the exception of the bronzes. There was, however, found here a fish-hook of this metal; the arrows are abundant there, and show the same workmanship; a little axe of serpentine has been found there, and several pieces of the same rock cut into wedges.

Settlement of Desor.—The accidents of this settlement are

truly remarkable. One would say that there had existed there a manufactory of earthenware, as everything which is not a vase appears excluded. It is from here that the only ornamented vases have been obtained, which have been discovered in the lake of Varese. The ornaments, as usual, consist of a series of straight lines disposed cheveronwise, or in little circles or semicircles, traced concentrically.

Settlement of Bardello.—This station, discovered by the Abbé Ranchet, presents a remarkable arrangement in the placing of the piles; there have been found there large cylindrical bones, split up, a mandible of the ox, manufactured bones, vases, etc.

Sig. Stoppani had scarcely terminated his explorations in the lake of Varese, when new investigations were initiated by Capt. A. Angeluni, director of the museum of artillery in the arsenal of Turin.* He also obtained happy results, especially as he found instruments in stone, as axes, arrowheads, etc.; among these last there were some made in the form of a heart, of really exquisite workmanship.† The palæo-ethnographical collection, contained in the museum of artillery, has been created by Sig. Angeluni. Either Sig. Stoppani or Sig. Angeluni found in the station of Bodio, and, as would appear, upon its very site, a considerable quantity of coin of the Roman epoch, the greater part of it silver, and belonging to the consular families.

We owe also to Sig. Stoppani the discovery in the lake of Garda‡ of five pile systems, two of which are at St. Felia, and three at the Isle of Lechi; but the discoveries were not to stop at the frontiers of Lombardy. Thanks to the activity of Dr. Lioy of Vicenza, we now know that certain regions beyond

---

* A. Angeluni, *Le Stazioni Lacunali del Lago di Varese;* Como, tip. Giorgielli.

† In the palæo-ethnographic collection of Valentino at Turin, there are, also, arrowheads of the same form and of an equally finished workmanship, coming from Monroe, in Pennsylvania. The collection of the Valentino is indebted for these arrowheads, as well as a collection of other arms in stone coming from the same region, to the liberality of M. A. Morlot.

‡ *Sulle antiche abitazione lacustri del Lago di Garda;* note, by A. Stoppani, *Atti della Soc. Ital. di Scienze Naturali,* vol. vi, p. 181, 1864.

the Adige are not less interesting a study, and not less rich in objects of high antiquity than those already known in Piedmont and Lombardy. The discoveries which he has made in the turbaries and caverns of Venetia are of a high importance.

Signor Lioy especially directed his investigations towards a marshy tract which borders the lake of Fimon, but which at some former time formed a part of the lake itself.* In a place named Pascolone he laid open a pile system, and having made an excavation, he first traversed 0·36 m. of turfy earth, and afterwards an argillaceous alluvial bed with shells of 0·40 m. in thickness, he then reached a bed of 0·30 m., entirely formed, of industrial, kitchen and table remains,—such things as ashes, chaff, split bones, crustaceans, tortoiseshell (*Emys lutaria*), nuts, acorns, etc. This bed contained no trace of metals, but there were an abundance of splinters of flint, among which was found an axe, and a hammer of stone,† large and small potteryware, some of which are ornamented, and a spindle-whirl. Under this bed he found the ancient bottom of the lake, rich in uniones, paludinæ, anodons, etc. The piles are planted in this bottom, and fastened together by transverse pieces, semicarbonised. Not far from the system of piles was found a great trunk of oak, hollowed out, and cut into the form of a canoe, one extremity of which terminates in a point. In his communication to the Society of Natural Sciences, which we have cited, Sig. Lioy mentions several other places where arrowheads and knives of flint have been discovered; these are Val di Barco, Val di Lonte, Monte Castello, Monte Grumi, Monte degli Schiavi, Padua, the neighbourhood of Treviso and San Vito da Tagliamento.

Sig. Lioy further explored certain caverns. In that of Colle di Mura, under stalagmites and a reddish *breccia* (*bréché*) which

---

* *Di una Stazione lacustre scoperta nel Lago di Fimon.* Communicazione del Socio Paolo Lioy, *Atti della Soc. Ital. di Scienze Naturali*, vol. vii, p. 167, (1864)

† This hammer, which was exhibited, and which we saw at the extraordinary meeting of the Society of Natural Sciences at Biella, appeared to be of porphyry, of a kind which ought to be known in the country. For further particulars, see "Le Abitazione Lacustri della Eta della pietra nel Vicentino di Paolo Lioy, con Tavoli." Venezia, Antonelli, 1865.

covered its bottom, he found, amidst cinders and charcoal, two arrowheads and other instruments, in silex; an earthen spindle-whirl, and a bone cut into a piercer, like those found by Lartet and Christy in the grottos of Eyzies, Laugerie, and Aurignac. In the cavern of Chiampo, at the same depth and in the same position, he found worked flints among the bones of a great bear.

From the north-east regions let us pass to those situated in a more southerly direction, where the study of high antiquity has made many proselytes. In 1863, Dr. T. Nicolucci communicated to the Academy of Naples a memoir, in which he has described some instruments and arms of stone,* axes, knives, arrowheads, etc., discovered in southern Italy, at Sora, Castelluccio, San Vicenzo in Val Roveto, Altamura, Ponte Corvo, Colle San Magno, Palezzolo, San Pietro in Curulis, Alvito, Monte San Giovanni, Civita Antica, Campoli, Luca, and Baljorano. The description of these arms of stone was a pretext simply to Sig. Nicolucci for a treatise upon the question of the ancient races, a subject into which, thanks to his talents and the study which he has given to it, he was able to import new ideas and facts. In 1864, he returned to this subject, by the publication of a fresh memoir.†

From the southern provinces we arrive at Tuscany, where, in addition to the discoveries made some years since at Cape Argentaro by the Marquis Strozzi, have been added those which Professor Cocchi has made public, in his letter addressed to M. Lartet.‡

We shall now pass over a long zone of country, either sterile or unexplored by the inquirers into high antiquity, and passing Genoa, we reach upon the western Riviera. Finale Marina, where we may signalise a cavern, which Messieurs

---

* "Di alcune armi ed utensili in pietra rinvenuti nelle provincie meridionali dell' Italia, e delle popolazioni ni tempi antestorici della peninsola Italiana." Memoria del Socio ordinario. G. Nicolucci, Napoli, 1863 estratta del vol. i degli Atti della R. Accademia.

† *La stirpe Ligure in Italia ni tempi antichi e ni moderni,* per G. Nicolucci, Napoli, 1864; estratto del vol. ii degli Atti della R. Accademia.

‡ *Sulla supposta antichita della Società Umana nella Italia centrale.* Firenze, 1864.

Issel and Perez visited in 1864, and in which they discovered human bones, marine shells, bones of ruminants manufactured, and fragments of pottery.* It has been observed that the human bones found in this cavern are calcined, and deeply hacked into with cutting instruments; and M. Issel inquires whether these hackings and the calcining do not afford indications of cannibalism among the ancient inhabitants of this part of Liguria.

We know that between Finale Marina and Nice there is a series of caverns where bones are found, as well as cinders and charcoal, and worked silex. Sig. Perez, whom we have cited above, discovered at Nice, in some wells which he had occasion to have executed at his château, a series of axes in variolite, spilite, spindle-whirls, and bones cut into piercers; axes of stone are, besides, common in the valley of Esteron, in the environs of Torretta, Giletta, Tadone, Pierre-a-feu, etc.

If we next traverse the maritime Alps, and descend into the hydrographical basin of the Po, we must notice the recent discovery of stone axes upon the hills, which go by the name of Langhe, lying on the right of the Tanaro; in the turbary of Avigliana, in which, up to the present time, nothing had been found, has been discovered a paalstab of bronze; and in Turin itself, at some meters depth, the blade of a dagger of bronze. Sig. Moro, in the turbary of Mercurago, and Sig. Telmoni, in one of those in the environs of Borgo Ticino, have discovered axes and arrowheads of stone, utensils in bronze, and worked wood. We must, however, signalise specially a fine canoe, formed of a hollowed-out pine-trunk, which they extracted from the bottom of a turbary in the neighbourhood of Ivrea, and which Dr. Gatta has been so kind as to give to the collection of the Valentino.

Descending the Po, and turning to the right bank of the river, we must notice several axes, two hammers, an arrowhead, and some *spindle-whirls*, in stone, as well as a paalstab of bronze, recently discovered on the Apennine of Piacenza,

---

* Issel, *una Caverna ossifera di Finale. Atti della Soc. Ital. di Sci. Naturali*, vol. vii, p. 173.

at Pessola, Bardi, Corviglio, Pellegrino, Gravago, Solignano, Campello, Vargi, etc.*

We shall terminate this rapid review at Parma. The environs of this town, as well as of Reggio and some parts of the Modenese, are an inexhaustible mine; and by all accounts, they are the part the best explored, best studied, and best described.

The most remarkable fact which these localities present is the existence of pile systems, under those deposits known as "*marnieras*," marl-beds or marl earths. Several of these marsh habitations have been discovered and described in 1863 and 1864; those of the Modenese by Canestrini, those of Reggio by Chierici, those of Parma and Pavullo, of Modena, by Pigorini. Among these pile systems, some are disposed in the form of a raft; and in the last few days (March 1865), Sig. Pigorini has discovered one at Fontanellato made of faggots, retained in position by a considerable number of piles.

We shall conclude by citing certain descriptive or synthetical works, published recently by Signori De Mortillet, Dal Pozzo and Garbiglietti. Such are, rapidly sketched, the discoveries either made or reported in Italy in the last few years. They have a real importance; and we may hope that if the Italians continue conscientiously to explore the soil of their country, they may contribute essentially to the progress of the study of high antiquity.

---

* Pallastrelli, *La Citta d' Umbria nell Apennino Piacentino*, Piacenza, 1864.

THE END.

PLATE I.

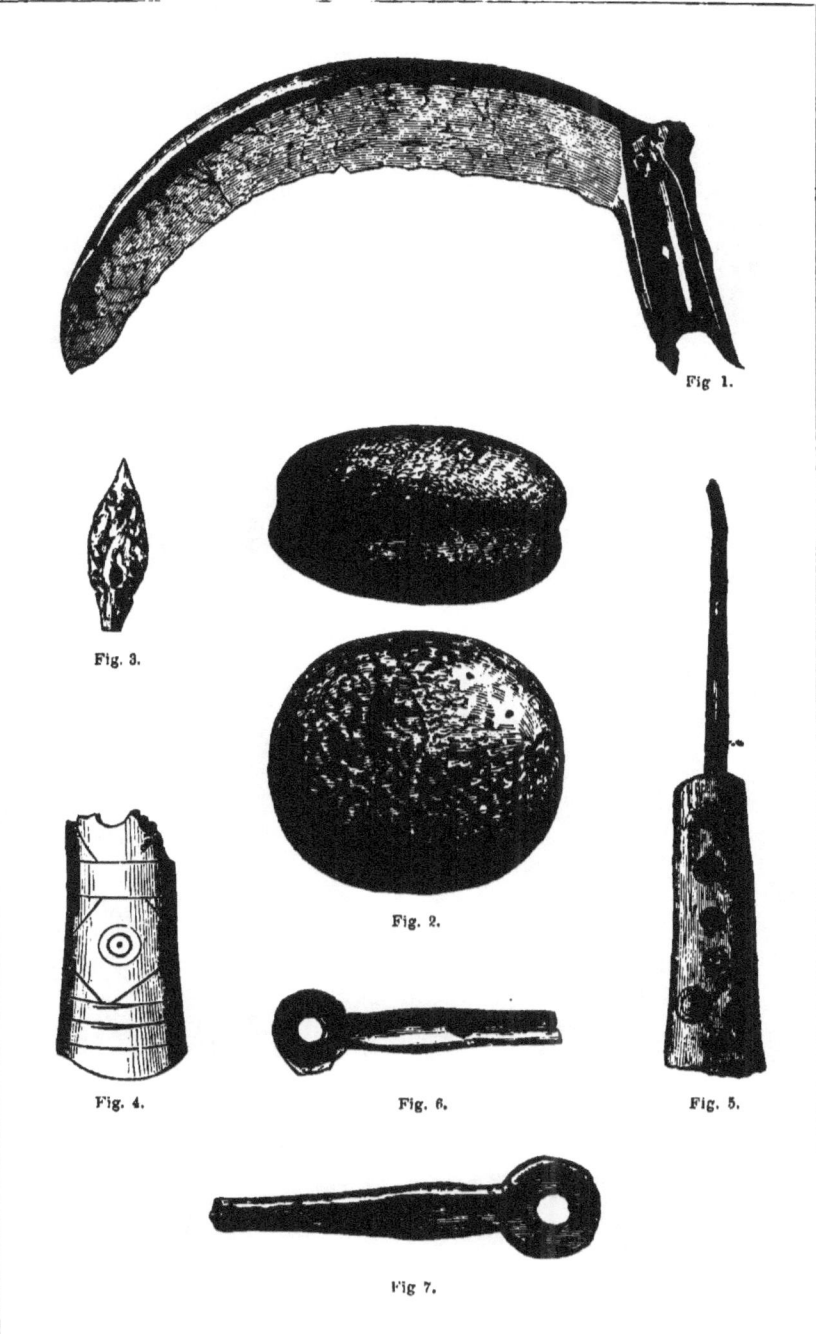

PLATE II.

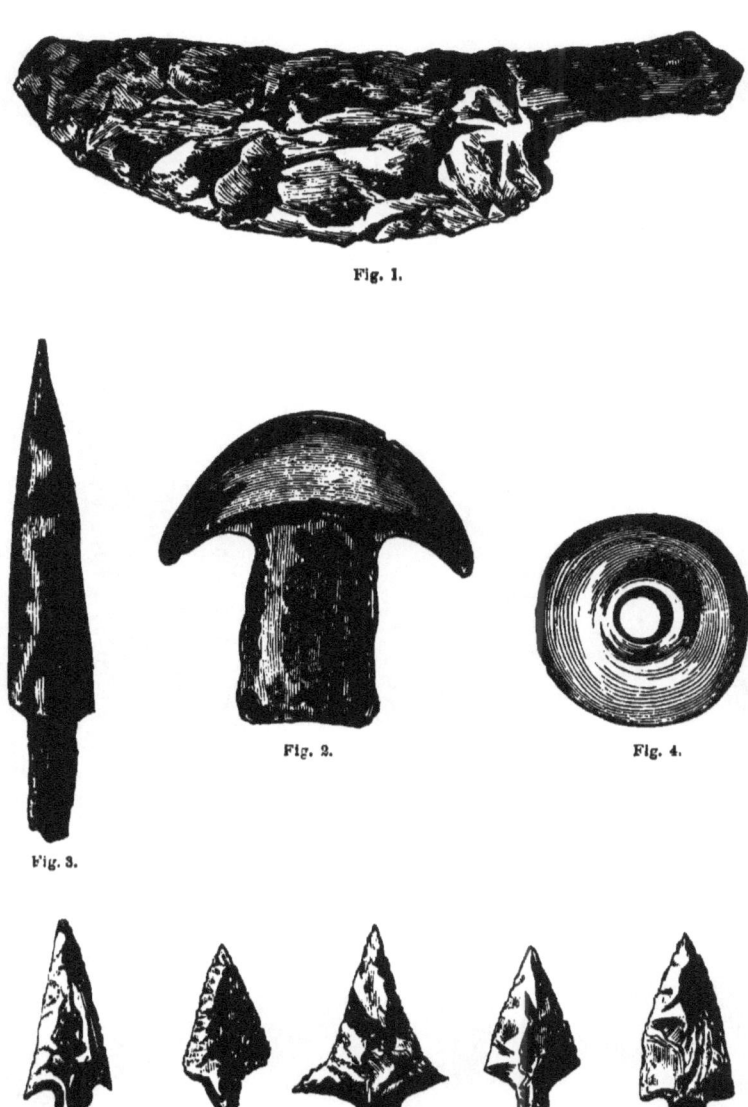

INDEX OF SUBJECTS.

Abbeville discoveries, 1
Ages of lacustrine habitations, 74, 118
Amber, 99
Amiens discoveries, 1
Analyses of earths, 11
Anani, 22, 89
Appended handles, 32
Apuan Alps, 90
Arrowheads, 6, 104
Ashes and carbon, 81
Aürochs, 65
Awls, 39

Badger, 65
Bardillo, 124
Beaver, 65
Birds, 69
Boar, wild, 51, 68
Bodio, 123
Boii, 12, 22, 89
Bone implements, 36
Bones, carved, 16
Bones, broken to extract marrow, 49
Borgo Ticino, 102
Bos, 57
Bos bison, 64
Bos brachyceros, 58
Bos frontosus, 65, 75
Bos primigenius, 58
Bos urus, 64
Brescia, 5
Bronze age, 6
Bronze utensils, 39
Bubalus, 65
Buffalo, 65
Bulldog, 52, 72
Butcher's dog, 52, 72

Canis, 51
Canoe, 102
Capra, 62
Cardium, 20
Castione, 36
Cat, 65
Cazzago, 123
Celt, 40
Cervus capreolus, 68
Cervus Elaphus, 66
Cervus megaceros, 66, 121
Chisel, 39
Civet-cat, 65

Comb, 37
Conventino di Castione, 18
Cornels, 104
Couteau-haches, 41, 96
Cumarola, 6
Cyclostoma elegans, 47

Desor, 123
Dog, 51
Domestic Mammalia, 49

Ears, 33
Earthenware, 20, 28
Elephas primigenius, 1, 120
Equus, 56
Etching, 108
Etrurians, 22, 89

Fallow-deer, 65
Fingones, 89
Fish-hooks, 39
Flints, manufactured, 2
Fox, 65
Frontal suture, 10
Fusajuole, 15, 44

Galli, 89
Goat, 51, 62
Grigioni, 53

Habitations, 24
Hazelnuts, 104
Hammers, 3, 40
Handles for pots, 31
Hare, 65
Hatchets, 3, 38
Hedgehog, 65
Horns, stags', 16, 17
Horse, 56
Human remains, 4, 6, 8, 9, 47
Hunting dog, 52
Hyæna spelæa, 1
Hypnum, 71

Insubres, 89
Isolino, 123
Ivrea, turbaries of, 113

Jade, 40
Jasper, 38, 40

Keller, 123
Kjökkenmöddings, 14
Knives, 39

Lacustrine habitations, 74
Lanceheads, 7
Lignite, turfy, 76
Lightning, protection from, 6
Lightning stones, 39
Lydian quartz, 3, 97

Majale, 74
Marl-beds, 12, 13, 77
Martin, 65
Mercurago, 102
Mica-schist, 3, 85
Mollusca, 70
Monte Tignoso caves, 4
Morainic turbaries, 99

Needles, 37, 104
Nephrite, 40
Norwegian houses, 25

Oberland, 53
Obsidian, 105, 106
Ollas, 35
Ophite, 40
Otter, 65
Ovis, 63, 64
Ox, 51, 57

Paalstab, 41, 94
Palisades, 20, 24
Panchina, 5
Paste of Celtic vases, 25
Pess, 37
Piedmont, 97
Pietra ollare, 34, 93
Piles, 103
Pins, 39
Piromaca, 6
Polecat, 65
Pottery, 25

Rat, 65
Rhinoceros leptorhinus, 66
Rhinoceros tichorhinus, 1
Roebuck, 65, 67

INDEX OF SUBJECTS.

Salso Maggiore, 21
San Martino, turbary of, 113
Sardinia, 97
Saussurite, 95, 97
Schiefer Kohle, 76
Scoriæ, 43
Secret of pottery, 34
Senones, 89
Serpentine, 7
Setter, 52
Sheep, 51, 63
Siamese, 53
Sicilian caves, 3
Skull, 9

Spindle-whirls, 15, 44, 97
Squirrel, 65
Stag, 51, 65, 66
Stone age, the, 2, 118
Sus scrofa, 53, 68
Suture, frontal, 10

Teeth, incisor, 5
Terra-cotta, 5
Terra mare, 23
Triticum turgidum, 71
Turbaries, dog of, 52, 99, 101

Umbrians, 89

Unio, 20, 21
Ursus arctos, 68
Ursus spelæus, 1
Urus, 76, 121
Utensils, 36

Vases, 29, 98
Vegetable remains, 71

Weasel, 65
Wheels, 38, 111
Wild animals, 50
Wires, 39

Zebu, 57

INDEX OF AUTHORS.

Angeluni, 124
Antoniono, 117, 118
Arnold, 91

Barbano, 114
Bertó, 18, 35, 68
Bertrand, 80
Bosis, De, 6
Bowring, 25
Brignoli, 9, 10

Cæsar, 76
Canestrini, 128
Capellini, 3
Cato, 92
Cavedoni, 6, 9, 12, 13, 78, 79
Cerchiari, 2
Chierici, 128
Cocchi, 126
Cocconi, Sig. Giacomo and Carlo, 36
Collomb, 119
Corbellini, 36
Contzen, 89, 91
Costa, 11
Curioni, 120

De Bosis, 95
De Mortillet, 94
Desor, 94, 118, 120, 122
Dionysius, 92
Doderlein, 12, 13, 16, 17, 96

Filippini, 5
Forel, 3

Gaddi, 11

Gastaldi, 122
Gatta, 114, 119, 127
Gatti, 6, 7, 9, 119
Ghiozzi, 79
Gibbon, 65
Gramizzi, 21, 70
Grassi, 16, 47, 95

Issel, 127

Keller, 85
Kohen, 91

Lartet, 1, 127
L'Haridon, Penguilly, 41
Linnæus, 74
Livius, Titus, 88, 125
Lopez, 14, 91
Luppi, 9

Manerini, 36
Mariotti, 78, 82, 84
Martini, 38
Mella, 115
Meneghini, 4
Merosi, 11
Micali, 87, 90
Morlot, 15, 124
Moro, 119, 127
Mortillet, 4, 66

Nepos, 89
Nicolucci, 96, 126
Niebuhr, 90, 91, 92

Orcasti, 94

Pullistrelli, 95, 128
Paul Warnefrid, 65

Perez, 127
Perthes, 5
Pigorini, 18, 23, 53, 92, 93, 118, 119, 128
Pliny, 89, 92

Ramazzini, 9
Ranchet, 124
Rezzi, 10
Ricci, 78
Rosa, 5
Rütimeyer, 51, 52, 55, 57, 58, 59, 66, 74, 76, 80

Sandri, 25
Scarabelli, 3, 95
Schweighauser, 91
Sismondi, 114
Stoppani, 123
Strabo, 25, 91
Strobel, 18, 23, 53, 92, 93, 118, 119
Strozzi, 3, 45, 126

Thomsen, 15
Thurnam, 11
Troyon, 15, 86
Truffi, 43

Ugolotti-Manarini, 21
Ugolotti-Romualdo, 21, 36

Venturi, 9, 12, 13, 14, 23, 78, 79
Vouga, 119

Welcker, 11

www.ingramcontent.com/pod-product-compliance
Lightning Source LLC
Chambersburg PA
CBHW030345170426
43202CB00010B/1246